THE DOOMSDAY CODE

God Is Warning Us Through The Bible

by Robert Fitzpatrick

Copyright © 2010 by Robert Fitzpatrick

The Doomsday Code
by Robert Fitzpatrick

Printed in the United States of America

ISBN 9781609571214

All rights reserved solely by the author. The author guarantees all contents are original and do not infringe upon the legal rights of any other person or work. No part of this book may be reproduced in any form without the permission of the author. The views expressed in this book are not necessarily those of the publisher.

Unless otherwise indicated, Bible quotations are taken from The King James Version.

www.xulonpress.com

But the end of all things is at hand: be ye therefore sober, and watch unto prayer.

1 Peter 4:7

This book is dedicated to my dear mother, Dorothy.

The author wishes to thank Mr. Harold Camping. His tireless teaching, writing and all around hard work at Family Radio for so many years have helped to make it possible for the true Gospel to go out all over the world.

Table of Contents

Preface xi

Introduction xiii

    Fulfilled Prophecy xiv

    Consistency xx

    Intriguing Scriptures xxv

Chapter 1: God's Creation 1

    The Creation 1

    The Spiritual Creation 5

    Ongoing Spiritual Creation 8

    A New Universe 10

Chapter 2: God's Book 13

    God Uses Irony 15

    God Didn't Record Everything 16

    Our Reaction to God's Words 17

    Only God's Words Can Save Us 21

    God's Words Condemn Us 23

    God's Gift to Mankind 24

    God Made It Difficult 25

    God Gives Us Understanding 26

    The Gospel Is Throughout 29

    How Should We Read the Bible? 31

    Translation Errors 33

    Progressive Revelation 39

Chapter 3: God's Law 43

    The Ten Commandments 43

The Whole Bible is a Book of Laws 45
Insight into the Ten Commandments 47
Purpose of the Law 50
Two Different Types of Laws and the Fourth Commandment 51
The Importance of God's Law 55
The Lord Jesus Fulfilled the Law 55
Other Changes to God's Laws 57
No Divorce 60
The Annual Feasts 60
The Sabbath Year 64
The Jubilee Year 65
Many Other Laws 67

Chapter 4: God's Problem 71
Man's Tendency 72
Man's Enemy 73
Man's Preoccupation 75
Man's Fallen Nature 76
Man and Music 77
Man and Alcoholic Beverages 79
Man and Government 81
Man is Special 82
God Grieves 83
The Fear of the Lord 85
The Saved 88
The Unsaved 91
Marriage and Children 93
Our Operating Manual 96

God's Instructions to Man 99
The Animals 99
Times and Seasons 102
God as Man 103

Chapter 5: God's Work 107
God's Future Work as Creator of a New Universe 107
God's Saving Work 108
Everyone Fits Into One of These Categories 115
God Intervenes 117
God's Work to Create and Preserve His Word 120
God Equips Us To Do His Will and Works Through Us 123
God's Intervention Affects Satan 126

Chapter 6: God's Calendar 131
Significance of Numbers in the Bible 133
The Calendar Patriarchs 141
About Carbon 14 Testing 144
Continuing With The Biblical Calendar 148
Synchronizing the Calendars 151
After the Fall of Judah 154
The Timeline Continues to the New Testament -
The Birth of the Lord 160
The Lord's Ministry Begins 165
The Crucifixion, the Resurrection, and Pentecost 167
A Prophecy Also Brings Us to 33 A.D. 170
The End of the Historical Timeline 172
Numerical Patterns in the Time Intervals 173

Chapter 7: God's Plan 177
God Has a Salvation Plan 177

The Adam-to-Noah Age 180
The Pre-Flood Age 182
The Noah-to-Moses Age 184
The Beginning of the Kingdom of Israel Until after the
Crucifixion 190
Pentecost and the Church Age 204
The Wheat and the Tares 207
God Finishes With the Earthly Church Near the End
of Time 209
The Number of the Beast and the End of the Church Age ... 213
A Scripture We Need To Examine Carefully:
Hebrews 10:25 214
A Shipwreck and the End of the Church Age 217
The Annual Feast Days and the Salvation Plan 218

Chapter 8: God's Calendar For Our Day 233
Time Confirmation in a Parable of the Fig Tree 233
1994 is a Key Year 235
Another Trail Leads to 1994 236
What Happened Before 1994? 238
The Great Tribulation Continues 239
The Final Feast of Tabernacles 243
A Review of the Path to Discovering the Dates 245
Possibilities for 1994 248
What Happens After 1994? 249
Numerical Patterns Provide Numerous Proofs 252
Further Confirmation of the Year 2011 as the Final Year 261
The Final Numerical Pattern: The Doomsday Code 263
Can We Know the Exact Hour Too? 266

Chapter 9: God's Wrath 271

The Earthquake 271
Dead Men's Bones 276
What About God's Mercy? 279
Why Us? 281
The Execution, The Shame and The Lost Inheritance 290
Some Want Punishment to Continue Forever! 295
About the Book of Revelation 296

Chapter 10: Now What? **303**
Why Should We Seek God's Mercy? 304
True Believers Have a Grave Responsibility 305
What Should An Unsaved Person Do Now? 308

Appendix 1: A Direct Derivation of the Rapture Date **313**

Appendix 2: Turning Back Our Calendar
　　　　　　To the Crucifixion **323**
An Anniversary Based on the Tropical Year 323
The Consistency of Spring 325
Israel's Calendar Was Tied to the Seasons 325
About Our Calendar 328
The 19 Year Cycle 329
Counting the Leap Years 331
Working Back From the Spring 2009 New Moon 334
When Was Nisan 1 in 33 A.D.? 335
The Date of the Crucifixion 336
It Was A Friday 336

Scripture Index **341**

Preface
The Doomsday Code

When we look at all the world's religions and the books or writings on which they are based - their holy books – we find that the Bible alone proves that it comes from God. The book you are holding is based on teachings from the Bible – the only source of truth.

You will read some astonishing things in *The Doomsday Code*. You may find this book to be deeply disturbing, even terrifying. It may mark a turning point in your life. On the other hand, you may decide to disregard it completely - even though it will give you proof that cannot be dismissed.

The ideas in this book, though shocking to those who are reading them for the first time, will be familiar to regular listeners of Family Radio. All the major themes and a great deal of everything else in this book are taken from the writings of Mr. Harold Camping of Family Radio and are discussed regularly in his teaching programs. These programs are aired over the Family Radio network of radio and television stations and reach millions of people all around the world.

In several books, Mr. Camping has explored the Biblical calendar, what the Bible reveals about end-time events - including the actual dates for these events, the significance of numbers in the Bible, the end of the church age, the nature of God's salvation, and other amazing teachings. These books are rich in detail and are truly Bible studies. In his books, Mr. Camping leads his readers into the forest - so to speak – to examine the trees under a magnifying glass. This book, on the other hand, will show you those trees through a pair of binoculars. In Mr. Camping's books, you will read many scriptures associated with each topic he covers.

In this book, you may read only a couple of scriptures that are considered typical or representative of the topic.

Therein lies the justification for this book. Mr. Camping's writings span decades and include many hundreds of pages. Also, his most recent books contain some teachings that are different from what is found in his earlier books. All of this is, of course, consistent with what we read in the Bible; for God has promised to give us understanding as we come close to the time of the end. Also, we should expect that our understanding of certain things will change as God opens our minds to His truth. The purpose of *The Doomsday Code* is to consolidate the major ideas you may read in Mr. Camping's books into one current book of intermediate size, thereby making these teachings more accessible to many persons.

The author wishes to thank Mr. Camping for his tireless work to spread the Gospel. God has truly blessed Mr. Camping with a keen analytical mind suited to uncovering the awesome truths hidden in the Bible, with an authoritative and pleasant speaking voice for teaching, and with a heart dedicated to His service. For over half a century, Mr. Camping and many other dedicated people have worked diligently to build Family Radio into a powerful tool for the Lord's use. This tool is now being used to warn the world that the time remaining is almost gone.

This book will prove to you how we may know the exact date of the world's end, why the earth will be destroyed, and how we know that it will happen. The source of this information is God's word, the Bible.

Introduction
The Doomsday Code

The terrifying information that will be presented in this book has come right out of the Bible, the word of God. You may be asking yourself why you should trust the Bible, or look to it for any kind of knowledge or wisdom. How can anyone know it really is the word of God? In a variety of ways, the Bible reveals itself to be beyond anything man could ever produce. In this Introduction, you will see why this is so. Later on, we will examine amazing mathematical proofs that come right out of the Bible. These proofs relate to end time dates of events in God's salvation plan; in the broader context, they also prove the Bible's Divine authorship.

God revealed the end time dates and mathematical proofs only a couple of years before the date on which the world shall end. For many centuries, however, the Bible has provided other proofs that it is indeed the word of God. Let's take a look at some of these longstanding proofs. The ones we will examine may be categorized as follows:

1. Fulfilled prophecy.
2. Consistency.
3. Some intriguing scriptures providing factual information about the real world, written long before the information was known to be true.

Fulfilled Prophecy

The Bible is a formidable book. The author's "giant print" King James Version is over 1800 pages long. Neither is it an easy book to understand. Theologians have been puzzling over some of its prophecies for hundreds of years. When we read the prophesies, we aren't going to find the type of clear, unambiguous language that people may expect when they hear the word "prophecy." We won't find, for instance, any paragraph in the Bible that says *"God will send His Son to be born in Bethlehem to a virgin in seven hundred years. He will suffer terribly to demonstrate His love for those whom God has chosen to save for eternal life. He will be crucified, die and rise from the dead. At the end of time, He will bring His wrath on all those who are not saved; but those who are saved He will take to live with Him forever."*

You won't find this composed paragraph in the Bible; but everything in it has been prophesied in the Bible and - except for that last sentence, which deals with the future - has already been fulfilled by the person who is known to history as Jesus Christ.

It has been estimated that about a third of the Bible deals with prophecy, and many of the prophecies are about Jesus Christ. Before we examine some of these amazing prophesies, we need to clarify a few things. The Bible has two main parts: the Old Testament and the New Testament. The Old Testament is considered "the book of the Jews." It consists of 39 books written over a long period of time. So far as we know, this period extends from about 1400 B.C. (before Christ, that is) to about 400 B.C.

The New Testament is considered by many to be "the book of the Christians." It consists of 27 books written approximately in the last half of the first century A.D. (Anno Domini in Latin, meaning "in the year of our Lord"). Most of the New Testament is about Jesus Christ but, incredibly, we find prophecies about the Lord Jesus even in the Old Testament. Jews do not accept the New Testament as being the word of God. Christians accept the Old Testament as being the word of God, but most Christians don't read it very much, preferring to concentrate on the New Testament if they read the Bible at all.

Introduction

When we talk about the Bible as being the word of God, we mean the whole Bible - all 66 books of the Old and New Testaments together. Those 66 books, and only those 66 books, are God's words - the Bible!

There are several well known English translations of the Bible. Despite the use of words such as "thee" and "thou" and many others that are not used in contemporary English, the King James Version of the Bible is recommended as an excellent translation. From the time of its completion in the year 1611 up to the present, it has been *the* Bible for generations of believers all over the world. The scriptures quoted in this book, unless otherwise indicated, are from the King James Version. Of course, the first number of each scriptural reference (following the name of the book) is the chapter of the book being quoted; the second number (after the colon) is the verse of that chapter. Let us now consider some of the prophecies that were fulfilled by the Lord Jesus.

The Book of Micah was written about 700 B.C. and tells us where He would be born. Micah 5:2 states:

But thou, Bethlehem Ephratah, *though* thou be little among the thousands of Judah, *yet* out of thee shall he come forth unto me *that is* to be ruler in Israel; whose goings forth *have been* from of old, from everlasting.

Although Micah 5:2 states that the subject will be a ruler in Israel, we need to understand that many times, when we read the word "Israel" in the Bible, it is not necessarily referring to the nation of Israel, or exclusively to the nation of Israel (ancient Israel or modern day Israel). In fact, many scriptures have more than one meaning. In many instances - as in Micah 5:2 - the word "Israel" must be understood to have a much broader meaning, applying to all those whom God chose to save. So we have here a prophecy that the Lord will come from Bethlehem.

We find the fulfillment of Micah 5:2 in the New Testament. For example, in Matthew 2:1-2, we read:

Now when Jesus was born in Bethlehem of Judea in the

days of Herod the king, behold, there came wise men from the east to Jerusalem, Saying, Where is he that is born King of the Jews? for we have seen his star in the east, and are come to worship him.

The Gospel According To Matthew is the first book of the New Testament, written hundreds of years after the book of Micah. Like the Book of Micah, the Book of Isaiah was written about 700 B.C. In Isaiah 7:14, we read another prophecy about the Lord's birth:

Therefore the Lord himself shall give you a sign; Behold, a virgin shall conceive, and bear a son, and shall call his name Immanuel.

The name Immanuel means "God with us." This prophecy was also fulfilled, as we read in Matthew 1:22-25:

Now all this was done, that it might be fulfilled which was spoken of the Lord by the prophet, saying, Behold, a virgin shall be with child, and shall bring forth a son, and they shall call his name Emmanuel, which being interpreted is, God with us. Then Joseph being raised from sleep did as the angel of the Lord had bidden him, and took unto him his wife: And knew her not till she had brought forth her firstborn son: and he called his name JESUS.

In the Book of Isaiah, beginning in chapter 52 and continuing into chapter 53, there are several scriptures that tell us about the Lord Jesus. In Isaiah 52:13, we read:

Behold, my servant shall deal prudently, he shall be exalted and extolled, and be very high.

Continuing in Isaiah 53:5 and through 53:7, about this same person we read:

But he *was* wounded for our transgressions, *he was* bruised for our iniquities: the chastisement of our peace *was* upon him; and with his stripes we are healed. All we like sheep have gone astray; we have turned every one

to his own way; and the LORD hath laid on him the iniquity of us all. He was oppressed, and he was afflicted, yet he opened not his mouth: he is brought as a lamb to the slaughter, and as a sheep before her shearers is dumb, so he openeth not his mouth.

Another part of our composed paragraph of prophecy can be found in Psalm 22, written by King David of ancient Israel roughly 300 years before the Book of Isaiah. In Psalm 22:14-16, we read:

I am poured out like water, and all my bones are out of joint: my heart is like wax; it is melted in the midst of my bowels. My strength is dried up like a potsherd; and my tongue cleaveth to my jaws; and thou hast brought me into the dust of death. For dogs have compassed me: the assembly of the wicked have enclosed me: they pierced my hands and my feet.

The Bible tells us that King David lived to old age and did not experience a terrible death, such as described in these and other verses of Psalm 22. You could argue that King David was writing these verses to help get himself through a difficult experience, that he was using exaggerated language in the process, and that this isn't a prophecy at all. You *could* say that; but the fact is that in the Gospels of the New Testament, written roughly 1,000 years after King David's Psalms were written, we find specific details of Christ's death on the cross fulfilling what we read in this Psalm. If you will refer back to these verses from Psalm 22 and the previously quoted passages of Isaiah chapter 53 when you read the Gospel accounts of the Lord's trial and execution, you will see the fulfillment of scripture.

The Gospels provide a factual account of events that had already transpired, and they show the fulfillment of Old Testament prophecies. In Luke 23, there is an account of the Lord Jesus when He was brought to Herod to be questioned. In Luke 23:9, we read,

Then he questioned with him in many words; but he answered him nothing.

So we see here the fulfillment of the prophecy in Isaiah 53:7 that "he openeth not his mouth." In Matthew 27:26, referring to Pilate's decision to release an infamous prisoner named Barabbas and to crucify Jesus, we read,

> **Then released he Barabbas unto them: and when he had scourged Jesus, he delivered *him* to be crucified.**

So when we read in the New Testament that the Lord was crucified, we have a fulfillment of Psalm 22:14-16. Another Psalm of David gives us a hint of Christ's resurrection. Psalm 16:10 tells us:

> **For thou wilt not leave my soul in hell; neither wilt thou suffer thine Holy One to see corruption.**

The word "suffer" here means "allow." From the New Testament, we know that the resurrection is a fulfillment of this prophecy of Psalm 16:10. Acts 2:29-30 refer to King David and then in the next verse, Acts 2:31 we read:

> **He seeing this before spake of the resurrection of Christ, that his soul was not left in hell, neither his flesh did see corruption.**

Now once again, the skeptic might say, "Well, whoever wrote the Book of Acts knew about Psalm 16 and just incorporated its verse 10 into Acts 2:31." In fact, it was actually only *after* the resurrection that many of the Old Testament scriptures were understood. Similarly, much of what the Lord Jesus told his disciples was not understood when He said it. It was only much later that many of His sayings were understood.

It should also be pointed out that, after the Lord rose from the dead, He was seen by *hundreds* of people. We learn this from 1 Corinthians 15:3-6, in which the apostle Paul is writing to a church at Corinth:

> **For I delivered unto you first of all that which I also received, how that Christ died for our sins according to the scriptures; And that he was buried, and that he rose again the third day according to the scriptures: And**

that he was seen of Cephas, then of the twelve: After that, he was seen of above five hundred brethren at once; of whom the greater part remain unto this present, but some are fallen asleep.

At the time those words were written (possibly about 55 A.D.), most of the hundreds of witnesses who had seen the risen Lord were still alive, although some of them had passed away (fallen asleep).

Now again, the skeptic might ask, "How do I really know anybody saw Christ walking around after the burial? Just writing it doesn't make it so." Well, I can think of a very powerful reason to accept this testimony from the Book of 1 Corinthians. Many of those first Christians were martyred. Of course, in a world of terrorism, such as our own, we know that many people are willing to die for their beliefs. But would someone be willing to die for a new "belief" that he knew to be a lie? Those first Christians were in a position to know the truth. They would not have been willing to die for their faith - in some cases a very painful death - unless they had *seen* the risen Lord.

The final sentence of our composed paragraph of prophecy has not yet been fulfilled. This sentence deals with the "Second Coming" - the return of the Lord at the end of the world. In 2 Thessalonians, Paul is writing to a church that has suffered persecution. In 2 Thessalonians 1:7-9, we read:

And to you who are troubled rest with us, when the Lord Jesus shall be revealed from heaven with his mighty angels, In flaming fire taking vengeance on them that know not God, and that obey not the gospel of our Lord Jesus Christ: Who shall be punished with everlasting destruction from the presence of the Lord, and from the glory of his power;

There are many scriptures that deal with the Lord's return and His judgment on the unsaved. Although these scriptures have not yet been fulfilled, there are many other prophetic scriptures that clearly have been fulfilled, and they are convincing proof that the Bible could only have been written by God.

Consistency

The consistency of the Bible is another proof that it is the word of God. Just think about the impossibility of such a long and complex book being produced by human effort, so that the final product provides consistent teaching and truth. Remember, the Bible was written over a period of about 1500 years. It was written down by over 30 men living in several different nations in basically two different languages - Hebrew for the Old Testament and Greek for the New Testament. The men who recorded it were very different from each other. Among them there were kings, priests, prophets, a herdsman - who was also a gatherer of fruit, a tax collector, a doctor, fishermen, and others.

How could each one have written his book or books to be consistent with everything else that had already been written? After all, it wasn't until about the year 1450 that the printing press was invented. Long before then, books - and when we refer to Biblical manuscripts we mean scrolls - were not readily available. It's most unlikely that each of those men had access to the scrolls of all the previously written books, so that he could consult the scrolls and study them.

There's no doubt that some scriptures are references to other scriptures that had been written earlier. At times, when we read the Bible, we may read the words "as it is written," or something similar. The apostle Paul, for example, was well educated and very familiar with the Old Testament. Familiarity with earlier scriptures, however, is by no means common to all the men God used to write the Bible. Also, we have reason to believe that many of the men credited with writing one or more of these books didn't even understand everything he had written down. Read, for example, what Daniel was told when he was given the last chapter of the book that bears his name. In Daniel 12:8, Daniel asks for an explanation of what he has heard:

And I heard, but I understood not: then said I, O my Lord, what *shall be* the end of these *things*?

In the following verse, Daniel gets an answer. Daniel 12:9 states:

Introduction

And he said, Go thy way, Daniel: for the words *are* closed up and sealed till the time of the end.

If we can't understand everything in the Bible - even the men God used to write it didn't understand it all - how then can we claim it is consistent? Well, the Bible has much to teach us; even though we don't understand everything in it, we can still see the common threads of God's handiwork running throughout the Bible. Furthermore, we understand more Biblical truth today than the ancients did, and this is in keeping with the principal of progressive revelation that we learn from the Bible: just as we read in Daniel 12:8 that Daniel didn't understand everything, we also learn from Daniel 12:9 that God will reveal more to those He has saved at "the time of the end." Later on, we will see that the Bible is consistent in this matter of progressive revelation.

As an example of the Bible's consistency, consider the commonly believed teaching in today's churches (and in many ministries outside those churches) that we can make a decision to accept Christ and that this is all we need do to guarantee our salvation. A scripture frequently quoted by those who accept this false gospel - one of many scriptures that you may come across when reading literature espousing that belief - is Revelation 3:20, in which the Lord Jesus says:

> **Behold, I stand at the door, and knock: if any man hear my voice, and open the door, I will come in to him, and will sup with him, and he with me.**

Another scripture often quoted in this regard is the well-known John 3:16:

> **For God so loved the world, that he gave his only begotten Son, that whosoever believeth in him should not perish, but have everlasting life.**

Admittedly, if you read these two scriptures (Revelation 3:20 and John 3:16) without considering everything else the Bible has to say, you can easily conclude "All I have to do is open that door and accept Jesus; all I have to do is believe in Him, and I will be saved." The Bible, however, doesn't teach this! Before we examine in greater detail this matter of whether or not anyone may control his own salvation, we need to take a detour.

When we read the Bible, we're not reading an ordinary book. Ordinarily, when we read anything other than the Bible, we will soon understand the writer's point of view - the position he's taking. So if a writer makes ten points clearly in support of a certain position, and then makes another point that is questionable, we'll think "Well, the point he's making in this book/article is obvious. He just didn't write that last item very well and it got past his editor too. We know what he means." But when we read the Bible, we can't do that. We can't interpret the Bible based on the weight of the evidence. We need to read *everything*. If there's just one scripture that says something that doesn't fit into our understanding of a dozen other scriptures, then our understanding of the dozen is not correct.

God's own words tell us that every word in the Bible counts. In Matthew 4:4, the Lord Jesus is quoting from the Old Testament:

> **But he answered and said, It is written, Man shall not live by bread alone, but by every word that proceedeth out of the mouth of God.**

And in 2 Timothy 3:16 we read:

> **All scripture *is* given by inspiration of God, and *is* profitable for doctrine, for reproof, for correction, for instruction in righteousness:**

Therefore, the Bible tells us that *all* scripture is from God. In Matthew 5:18, in which the Lord Jesus is discussing the scriptures we learn the importance of every word:

> **For verily I say unto you, Till heaven and earth pass, one jot or one tittle shall in no wise pass from the law, till all be fulfilled.**

Yes, when God tells us in writing that He's going to do something, He "dots the i's" and "crosses the t's." Therefore, the Bible itself clearly tells us that we need to read everything in it, because it's all important - it's all the word of God.

Getting back to what the Bible really teaches about God's salvation plan - we know that there is nothing that anyone can do to get himself saved because the scriptures agree on this. Only this way of understanding satisfies the requirements imposed by all the scriptures dealing with salvation. For example, in the Book of Ephesians, we learn that God chose those He wanted to save long before they were born. In Ephesians 1:4-6, we read:

> **According as he hath chosen us in him before the foundation of the world, that we should be holy and without blame before him in love: Having predestinated us unto the adoption of children by Jesus Christ to himself, according to the good pleasure of his will, To the praise of the glory of his grace, wherein he hath made us accepted in the beloved.**

"Well," you might say, "God knew which individuals would accept Him of their own free will, so he chose to save those people." This is one way these verses of Ephesian 1:4-6 are commonly reconciled with other verses that *seem* to teach that we make a contribution to our own salvation - for example, by accepting Jesus, or by attending a mass every week without fail, or getting baptized in water, or by just believing in Him, or by doing anything else. However, that way of understanding cannot be correct because from other scriptures we learn that the process of salvation - from start to finish - is entirely the work of God.

For example, in the Book of Romans we learn something of our true condition apart from God's mercy. In Romans 3:10-11, we read:

> **As it is written, There is none righteous, no, not one: There is none that understandeth, there is none that seeketh after God.**

On our own we won't seek God, at least not with our heart truly set on Him; but if an individual is one of God's elect, God will begin to work in that person's life. In John 6:44, we read something about this process:

> **No man can come to me, except the Father which hath sent me draw him: and I will raise him up at the last day.**

We don't have any control over any step in this process. The timing is completely in God's hands, and we have to wait on Him, as we read in Isaiah 40:31:

> **But they that wait upon the LORD shall renew *their* strength; they shall mount up with wings as eagles; they shall run, and not be weary; *and* they shall walk, and not faint.**

Even those whom God has chosen to save - the ones who will be caught up in the air with the Lord at His return in their new, glorified, bodies - even they must wait for the Lord to perform in them the good work He will do. And this work is entirely God's doing. Ephesians 2:8-9 state:

> **For by grace are ye saved through faith; and that not of yourselves: *it is* the gift of God: Not of works, lest any man should boast.**

Many verses speak of this saving kind of faith; but, as we see from these verses, even this faith is a gift of God. There is no kind or amount of work that we do on our own - and faith is considered a work - that can save anyone. There is nothing anyone can say, do or even think that will save him. This is a hard truth to accept, and many people do not accept it. It is hard to accept because we want to be in control of our own destiny. When we learn that we aren't in control, our pride may be offended; we may experience fear for ourselves and our loved ones, and sorrow over those whom we suspect have been lost for eternity. Nevertheless, this is what the Bible consistently teaches about salvation, even though it may not always be obvious. This very consistency in what the Bible teaches about salvation and other matters is proof that the Bible is the word of God.

Intriguing Scriptures

In the Bible, there are several intriguing scriptures that reveal truths about this world - truths that were not commonly known until long after they were written in the Bible. An example of one of these scriptures is Isaiah 40:22 where, referring to God, we read:

> *It is* **he that sitteth upon the circle of the earth, and the inhabitants thereof** *are* **as grasshoppers; that stretcheth out the heavens as a curtain, and spreadeth them out as a tent to dwell in:**

So here we have a clear statement in the Book of Isaiah that the earth is round. If you consult an encyclopedia, you will probably find that the ancient Greeks - just around 500 B.C. or so - knew that the world is a sphere. The Book of Isaiah was written about 200 years earlier than that. Actually, there is a scripture from the Book of Job that also indicates that the earth is round, although it doesn't say it so clearly in our King James translation. Scholars think that the Book of Job may have been written many centuries earlier than the Book of Isaiah.

The Book of Leviticus contains many laws given to ancient Israel. One of these laws is a prohibition against eating blood. This Biblical law, by the way, is still in effect today. In the part of the Bible dealing with this, we read, in Leviticus 17:11:

> **For the life of the flesh** *is* **in the blood: and I have given it to you upon the altar to make an atonement for your souls: for it** *is* **the blood** *that* **maketh an atonement for the soul.**

This scripture, like so many others, points to man's need for a Savior; but look again at the first part of that scripture: "the life of the flesh is in the blood." It's been a while since the medical profession has dropped bloodletting as a common treatment for many ailments. However, when we realize that this part of the Bible was written about 1400 B.C. and that the practice continued in the United States into the Civil War era, we can only wish that doctors had really believed this scripture. Many men and women, including George Washington in 1799, died at the hands of their doctors because bloodletting was used to treat them. Leviticus 17:11 is, therefore, a striking example of life-saving knowledge that God revealed to mankind in His word.

If you investigate the first uses of quarantine as a way to control outbreaks of infectious disease, you may learn that it wasn't until the 1400's (A.D.) that a formal system was established. That's when Venice required ships to lay at anchor for 40 days before landing, following the devastation of the Black Death. Earlier than that, in 549 A.D. the emperor Justinian enacted a law meant to isolate anyone from an area affected by bubonic plague. In the Book of Leviticus, however, we find the law of quarantine given to ancient Israel approximately 2,000 years before Justinian's law. The thirteenth chapter of Leviticus deals with leprosy, and in Leviticus 13:46 we read:

All the days wherein the plague *shall be* in him he shall be defiled; he *is* unclean: he shall dwell alone; without the camp *shall* his habitation *be*.

God seems to be giving us a picture of the awfulness of sin in all the scriptures about leprosy; but the fact is that in Leviticus 13:46 we have a statement of the law of quarantine.

As you read this book, you will see further proof that the Bible is indeed the word of God and that we had better take heed of its warnings.

Chapter 1

God's Creation

We know that the universe is absolutely beyond comprehension in its vastness. Our great solar system - consisting of the sun, our earth and all the other planets that revolve around the sun - is just a minuscule part of the galaxy containing it. That galaxy, the Milky Way, contains *billions* of stars. Its dimensions are measured in thousands of light years (a light year is the distance traveled in one year when traveling at the speed of light); and the Milky Way is only one of *billions* of galaxies in the universe. Whose mind can grasp this immensity? The Bible reveals that God is the Creator of this astounding universe.

The Creation

The first two chapters of Genesis tell us that God created the universe, all plants and all creatures in six literal days. We read in Genesis 1:31:

> And God saw every thing that he had made, and, behold, *it was* very good. And the evening and the morning were the sixth day.

The universe was fully operational immediately after God created it: starlight from distant galaxies was well on its way to earth; animals known from our study of nature to train their young in some aspect of survival in the wild - these animals were already fully equipped to survive; and the first man was created not as a baby, but as an adult who could understand the world and converse with God.

It's really not surprising that so many people refuse to believe it happened that way. After all, we are taught in school that the universe resulted from a massive explosion billions of years ago, that life on the earth developed by chance, and that from that earliest, simple life-form, everything else - including mankind - has evolved. However, if you have learned anything about science, it should occur to you that this "big- bang" / evolution explanation is plain nonsense.

Order doesn't result from an explosion; neither did the placement of the earth relative to the sun. If the earth were just a *little* bit closer to the sun, our planet would be too hot to sustain life; if it were just a *little* further away, we would all freeze. And the sun itself is remarkable as a star in its consistent energy generation. If it varied in its output as so many other stars do, temperatures on the earth would soon go outside the bounds required for human life. The sun obviously didn't emerge from an explosion - it was designed to do its job.

What about those first simple life forms that supposedly came out of some primordial "soup" that had accumulated on the earth? We now know that the simplest life forms aren't so simple after all. With the powerful magnification of currently available instruments, scientists have learned that even single-celled organisms are extremely complex little machines. There's about as much chance of one of them just happening into existence as there is of a running automobile just happening to come together after a storm blows over a junkyard.

The idea that evolution is the explanation for the existence of man or any other creature would be laughable if it hadn't been responsible for so much misery. The interdependency of species and the observed tendency for natural processes to proceed toward a state of greater disorder are just two of a great many arguments against it. In fact, there is a great deal we learn from science that supports what we read in the Bible about origins. We can trust the Bible's creation account.

It isn't just in the Old Testament that we learn that God created the universe. For example, in John 1:1-3, we read:

Chapter 1 God's Creation

> **In the beginning was the Word, and the Word was with God, and the Word was God. The same was in the beginning with God. All things were made by him; and without him was not any thing made that was made.**

God is also revealed as the Creator in Acts 17:24-25, where the apostle Paul is seen explaining to the men of Athens:

> **God that made the world and all things therein, seeing that he is Lord of heaven and earth, dwelleth not in temples made with hands; Neither is worshipped with men's hands, as though he needed any thing, seeing he giveth to all life, and breath, and all things;**

Some people have thought of God as just sitting back and letting the universe go on its own after He created it, as if the universe were a great clock that had been wound up to run for the time allotted to it. The Bible tells us that this isn't the case at all. God keeps the universe on track, as we read in Hebrews 1:3:

> **Who being the brightness of *his* glory, and the express image of his person, and upholding all things by the word of his power, when he had by himself purged our sins, sat down on the right hand of the Majesty on high;**

The "Who" of this scripture is referring to the Lord Jesus, the Son of God, who is upholding all things by the word of his power. The Bible also indicates God's sustaining power in Acts 17:28, where we read:

> **For in him we live, and move, and have our being; as certain also of your own poets have said, For we are also his offspring.**

Those words were spoken by the apostle Paul to the men of Athens, when he proclaimed to them the one true God. We're also given a hint of God's sustaining power in a message God delivered to king Belshazzar through the prophet Daniel. In Daniel 5:23, we read:

> **But hast lifted up thyself against the Lord of heaven; and they have brought the vessels of his house before**

> thee, and thou, and thy lords, thy wives, and thy concubines, have drunk wine in them; and thou hast praised the gods of silver, and gold, of brass, iron, wood, and stone, which see not, nor hear, nor know: and the God in whose hand thy breath *is*, and whose *are* all thy ways, hast thou not glorified:

Notice that this scripture is telling us that the king's breath and all his ways are in God's hand. The king's life - and of course this also applies to each one of us - could continue only as long as God allowed it; neither could it end unless God allowed it.

God truly sustains the world; but there have been times when he has brought judgment against it. After the first man sinned, God did something that I'm sure He took no pleasure in doing. We learn that God had planted a garden, and that's where He put Adam, the first man. In Genesis 2:15-17, we read about this:

> **And the LORD God took the man, and put him into the garden of Eden to dress it and to keep it. And the LORD God commanded the man, saying, Of every tree of the garden thou mayest freely eat: But of the tree of the knowledge of good and evil, thou shalt not eat of it: for in the day that thou eatest thereof thou shalt surely die.**

God obviously created the language He was using to speak with Adam, and gave Adam the ability to use language. Adam understood God's command; but he and his wife Eve, the first woman (another creation of God, as was the institution of marriage), did not obey. Eve ate of the forbidden tree after being tempted by Satan, and then Adam ate of it too. Consequently, God drove them from the garden.

As a result of their sin, God placed a curse on the earth. We read this in Genesis 3:17-19:

> **And unto Adam he said, Because thou hast hearkened unto the voice of thy wife, and hast eaten of the tree, of which I commanded thee, saying, Thou shalt not eat of**

it: cursed *is* the ground for thy sake; in sorrow shalt thou eat *of* it all the days of thy life; Thorns also and thistles shall it bring forth to thee; and thou shalt eat the herb of the field; In the sweat of thy face shalt thou eat bread, till thou return unto the ground; for out of it wast thou taken: for dust thou *art*, and unto dust shalt thou return.

I'm sure it grieved God to do it. He had given man a perfect world; but the world would now become a harsh, hostile, environment. Why did God do this? Perhaps because man was no longer compatible with the earth as it had been created. Fallen mankind would have to live in a cursed world.

When God warned Adam not to eat of that one particular tree (Genesis 2:17), He said "for in the day that thou eatest therof thou shalt surely die." Yet we read in Genesis 5 that Adam went on to live for hundreds of years after this. Adam obviously didn't die that day, at least not physically. How then can we understand this warning to Adam?

The Spiritual Creation

God's creation is not restricted to the material world. In Colossians 1:16, we read about the Lord Jesus and the creation:

For by him were all things created, that are in heaven, and that are in earth, visible and invisible, whether *they be* thrones, or dominions, or principalities, or powers: all things were created by him, and for him:

Apparently, the creation consists of much more than what we see. For example, the Bible reveals that God created many spiritual beings called angels. They were created to serve Him. Psalm 148:2 has one of many references to angels:

Praise ye him, all his angels: praise ye him, all his hosts.

Luke 15:10, in which the Lord Jesus is speaking to His disciples, has another reference to angels:

> **Likewise, I say unto you, there is joy in the presence of the angels of God over one sinner that repenteth.**

In many of the Bible's references to angels, we learn that they are very powerful beings. For example, in Matthew 28 we read about the women at the tomb of the Lord Jesus on the Sunday following the Crucifixion. In Matthew 28:2-4, we read:

> **And, behold, there was a great earthquake: for the angel of the Lord descended from heaven, and came and rolled back the stone from the door, and sat upon it. His countenance was like lightning, and his raiment white as snow: And for fear of him the keepers did shake, and became as dead *men*.**

In addition to scriptures about angels, there are other references to God's spiritual creation. We get a glimpse of God's throne in the first chapter of Ezekiel. In Ezekiel 1:26, we read:

> **And above the firmament that *was* over their heads *was* the likeness of a throne, as the appearance of a sapphire stone: and upon the likeness of the throne *was* the likeness as the appearance of a man above upon it.**

The apostle John also had a vision of God's throne. In Revelation 4:2-3 we read:

> **And immediately I was in the spirit; and, behold, a throne was set in heaven, and *one* sat on the throne. And he that sat was to look upon like a jasper and a sardine stone: and *there was* a rainbow round about the throne, in sight like unto an emerald.**

Just as we learn from the Bible that God is a Spirit, I think we may safely say that God's throne is part of His spiritual creation.

There is something else that the Bible reveals about God's spiritual creation. We read in Genesis 1:27:

> **So God created man in his *own* image, in the image of God created he him; male and female created he them.**

Man was created to be totally different than all of God's other creatures. He was created in the image of God; but God is a Spirit. From this scripture and many others that deal with man's nature and his situation in eternity, we get a picture showing us that God holds man responsible for his sins. God will save some to be with Him for eternity. Those whom He does not save are under God's wrath and are destined for destruction: they will die and never live again. We read something about this in John 3:36:

> **He that believeth on the Son hath everlasting life: and he that believeth not the Son shall not see life; but the wrath of God abideth on him.**

Another scripture pertaining to the fate of the unsaved is Revelation 14:11:

> **And the smoke of their torment ascendeth up for ever and ever: and they have no rest day nor night, who worship the beast and his image, and whosoever receiveth the mark of his name.**

Although it is not obvious from this scripture that the unsaved will suffer annihilation rather than punishment without end in a place called hell, when we read the Bible carefully we learn that annihilation is in fact the punishment God has decreed.

God warned Adam not to eat of the tree of the knowledge of good and evil, saying "in the day that thou eatest therof thou shalt surely die." Certainly, as a result of their disobedience Adam and Eve were banished from the garden and eventually suffered physical death. That much is clear from the Genesis account. Something else, however, happened to them that day when they disobeyed God's command: Adam and Eve had been created with the potential to live for eternity; they lost this as a result of their sin - they died spiritually. This was the death God had spoken about in His warning to Adam. They had come under the wrath of God and they would be subject to judgment at the end of time. The Bible refers to this judgment as a second death, as we read in Revelation 21:8:

> **But the fearful, and unbelieving, and the abominable, and murderers, and whoremongers, and sorcerers, and**

idolaters, and all liars, shall have their part in the lake which burneth with fire and brimstone: which is the second death.

The second death marks the destruction of any remains of those who have died unsaved; there is no hope for these people because they will never live again. In a remark to one of His disciples, the Lord Jesus referred to this condition of being subject to God's wrath. We read about this in Matthew 8:21-22:

> And another of his disciples said unto him, Lord, suffer me first to go and bury my father. But Jesus said unto him, Follow me; and let the dead bury their dead.

In saying "let the dead bury their dead," the Lord referred to a spiritual problem common to every person who has ever lived - unless that person has been saved by God.

Ongoing Spiritual Creation

The Bible reveals that God's work of spiritual creation is ongoing. Whenever God saves someone, He gives that person a new soul. It is the resurrection of the spiritually dead. This is a mighty act of creation that only God Himself is capable of doing. It is rebirth, on the spiritual plane of existence. The Lord Jesus explained something about this in a discussion with a man named Nicodemus, as we read in John 3:6-8:

> That which is born of the flesh is flesh; and that which is born of the Spirit is spirit. Marvel not that I said unto thee, Ye must be born again. The wind bloweth where it listeth, and thou hearest the sound thereof, but canst not tell whence it cometh, and whither it goeth: so is everyone that is born of the Spirit.

There is nothing to see when this spiritual rebirth occurs. It is, however, the gift of everlasting life, as we read in John 5:24:

> Verily, verily, I say unto you, He that heareth my word and believeth on him that sent me, hath everlasting life.

and shall not come into condemnation; but is passed from death unto life.

We cannot see a person's new soul - it is invisible, like the wind. Once God has done this for someone, however, that person has undergone a fundamental change and will never be the same. He or she has become someone who wants to do God's will. Of course, there will still be struggles with sin. We read something about this struggle in Romans 7:22-23:

> For I delight in the law of God after the inward man: But I see another law in my members, warring against the law of my mind, and bringing me into captivity to the law of sin which is in my members.

Even someone who has been born again is still vulnerable to temptation to sin. His body is still infected by sin. In his new soul, however, he no longer wants to sin. He is now a child of God. He has been plucked out of the kingdom of Satan and brought into the kingdom of God. As a result, that individual has a new heart. At first, it may not even be noticeable to anyone, not even to the person who has been saved; but in time, the person will notice more and more that he wants to be obedient to God's laws. God tells us about this in Ezekiel 36:27:

> And I will put my spirit within you, and *cause* you to walk in my statutes, and ye shall keep my judgments, and do *them*.

This spiritual change permits a person to begin to live as God intended, avoiding the world's sinful activities, as we read in 1 Peter 2:9-10:

> But ye *are* a chosen generation, a royal priesthood, an holy nation, a peculiar people; that ye should shew forth the praises of him who hath called you out of darkness into his marvelous light: Which in time past *were* not a people, but *are* now the people of God: which had not obtained mercy, but now have obtained mercy.

As we learn from this scripture, someone who has become a child of God - one of the "people of God" - has obtained mercy from God. The threat of destruction no longer hangs over that person because the Lord Jesus has paid for his sins. This is how the apostle Paul knew we need not fear death, because physical death brings us to be with the Lord, as we read in 2 Corinthians 5:8:

> **We are confident, *I say*, and willing rather to be absent from the body, and to be present with the Lord.**

Not only has salvation eliminated the threat of eternal death; it has also given the saved person the promise of an eternity with God Himself. God's children will be with Him, living in happiness that we can't imagine, as we read in Ephesians 2:4-7:

> **But God, who is rich in mercy, for his great love wherewith he loved us, Even when we were dead in sins, hath quickened us together with Christ, (by grace ye are saved;) And hath raised *us* up together, and made *us* sit together in heavenly *places* in Christ Jesus: That in the ages to come he might shew the exceeding riches of his grace in his kindness toward us through Christ Jesus.**

A New Universe

Those of God's children who have passed away throughout history have gone to be with the Lord in their spirit. Shortly before the end of the world, however, God will resurrect them and give each a new body - a body that will be glorious for all eternity.

God's word tells us in no uncertain terms that this world *will* eventually come to an end. It won't only be the end of our planet - it will be the end of the entire universe. This event won't occur millions of years from now because, for example, the sun has consumed itself; and it won't happen because man has finally destroyed himself as a result of war. It *will* happen when God has determined it will happen, and that will be soon. God will destroy the universe when the time for it to exist has ended, as we read in 2 Peter 3:10:

> But the day of the Lord will come as a thief in the night; in the which the heavens shall pass away with a great noise, and the elements shall melt with fervent heat, the earth also and the works that are therein shall be burned up.

Incredibly, along with everything else it reveals about God's plans for mankind, the Bible also tells us a great deal about the timing of major events in God's plan. So even though we read that the last day will come "as a thief in the night," elsewhere in the Bible we learn that it is only the unsaved who will be taken by surprise, as we read in 1 Thessalonians 5:3-4:

> For when they shall say, Peace and safety; then sudden destruction cometh upon them, as travail upon a woman with child; and they shall not escape. But ye, brethren, are not in darkness, that that day should overtake you as a thief.

In fact, the Bible proves that we are extremely close to the end of the world. God promises us a new and glorious creation after He has destroyed this present universe. We read of this in Revelation 21:1:

> And I saw a new heaven and a new earth: for the first heaven and the first earth were passed away; and there was no more sea.

God will create a new heaven and a new earth. The new earth will be as beautiful as the present earth originally was, when it was first created. It will have none of the causes for our suffering that our sin-cursed world has, as we read in Revelation 21:4:

> And God shall wipe away all tears from their eyes; and there shall be no more death, neither sorrow, nor crying, neither shall there be any more pain: for the former things are passed away.

Chapter 2

God's Book

God is an author because He wrote the Bible - it is His book. Of course, it wasn't actually His hand that put the words down on the writing material. He used men to do that; but just as surely as each man held a writing instrument in his hand, each of these men was himself an instrument in God's hands to write His words.

God is not generally credited as being an author. Most people will probably tell you that men wrote the Bible. Some may say that men were inspired by God to write in their own words whatever it was that God wanted generally to tell us. The Bible, however, isn't what God generally wanted to tell us - it is a book in His own words, *exactly*. The Bible tells us how we actually got the Bible. In Jeremiah 36:1-2, we read:

> **And it came to pass in the fourth year of Jehoiakim the son of Josiah king of Judah,** *that* **this word came unto Jeremiah from the LORD, saying, Take thee a roll of a book, and write therein all the words that I have spoken unto thee against Israel, and against Judah, and against all the nations, from the day I spake unto thee, from the days of Josiah, even unto this day.**

(Note that italicized words appearing in the scriptural quotations were added by the translators as they thought necessary to convey the meaning and were not part of the original, inspired text.) In many parts of the Bible, the man to whom God was dictating knew that he was writing God's words. For example, when Jeremiah wrote the words "Take thee a roll of a book," he knew that these were God's words. God had previously spoken these very words to him, and Jeremiah was quoting God. When we read a scripture such as 2 Thessalonians 3:17, however, we might not be so certain that the words are God's own words:

The salutation of Paul with mine own hand, which is the token in every epistle: so I write.

Did God write those words? Yes, He did. Even though in these words Paul is pointing out his style of handwriting to those who would read his epistle, nevertheless the words came from God.

How can it be that God is responsible for Paul's words about his own handwriting? If you were to sit down and write a letter to your friend, the words you wrote would be the words that came into your mind as you composed the letter. They would be your own words. Suppose, however, that for His own purpose God wanted this to be a special letter to your friend. He could put into your mind the words He wanted you to write, yet you would think they were your own words. Similarly, Paul's hand wrote down those words as they came into his mind when they were written for the very first time; but the words belong to God.

On one occasion, when the Lord Jesus was telling His disciples that they would face persecution, He said something that shows us how God wrote the Bible. In Matthew 10:17-20, we read:

But beware of men: for they will deliver you up to the councils, and they will scourge you in their synagogues: And ye shall be brought before governors and kings for my sake, for a testimony against them and the Gentiles. But when they deliver you up, take no thought how or what ye shall speak: for it shall be given you in that same hour what ye shall speak. For it is not ye that speak, but the Spirit of your Father which speaketh in you.

In the Acts of the Apostles, we read of several occasions when the Jews seized one or more of the apostles - as the Lord had warned them. In these accounts, we read the words spoken by the apostles.

Those words were given to the apostles just as the Lord Jesus said they would be; and those words were the inspired words of God, just like everything else we read in the Bible.

God Uses Irony

As you read some parts of the Bible, you may notice something interesting about God's writing: He uses irony. Your dictionary may define it as something like this: "expression in which the intended meaning of the words is the opposite of their usual sense." The intended meaning of a scripture will be its spiritual meaning: that's what God is teaching us. Sometimes the spiritual meaning is extremely difficult to understand, especially because the apparent meaning contradicts it. In Matthew 19:16-17, we see an example of irony in the Lord's response to a question He was asked:

> **And, behold, one came and said unto him, Good Master, what good thing shall I do, that I may have eternal life? And he said unto him, Why callest thou me good?** *there is* **none good but one,** *that is,* **God: but if thou wilt enter into life, keep the commandments.**

The Lord's words in this scripture prove that we must read the Bible so very carefully in order to avoid falling into traps of misunderstanding. The Lord appears to be telling His questioner that, because only God is good, He should not be addressed as "good." According to the usual sense of that response, the meaning is: "I'm not God." Of course, we know from many other scriptures that the Lord Jesus was and is God, and that He never ceased to be God.

The Lord's response also seems to imply that someone may be able to earn eternal life by keeping the commandments. We know now that we can't earn eternal life. The Lord would, however, grant it to someone who keeps the commandments perfectly; but no one can do that. Even if someone could at some time in his life begin to keep the commandments perfectly, there is still the matter of all the sins he or she committed before then.

There is no remedy for those sins, unless the Lord Jesus has paid for them.

We also see irony in a question asked by the woman at the well in a city called Sychar. She was a Samaritan woman who had a discussion with the Lord Jesus. In John 4:12, we read this question:

> **Art thou greater than our father Jacob, which gave us the well, and drank thereof himself, and his children, and his cattle?**

Her question was sarcastic, but it was certainly ironic because she addressed it to the One who indeed was greater than Jacob, or any other human to ever live.

God Didn't Record Everything

As we read the Bible, we should realize that there were other writings - possibly very many other writings - dealing with times, people, places and events we read about in the Bible. Those other writings didn't make it into the Bible because God didn't include them. The Bible's 66 books are the only ones God chose to preserve for the ages, and they were written and preserved for us in the original languages exactly the way God wanted them to be. The opening verse of The Gospel According to Luke hints at some other writings from the New Testament period. In Luke 1:1-4, we read:

> **Forasmuch as many have taken in hand to set forth in order a declaration of those things which are most surely believed among us, Even as they delivered them unto us, which from the beginning were eyewitnesses, and ministers of the word; It seemed good to me also, having had perfect understanding of all things from the very first, to write unto thee in order, most excellent Theophilus, That thou mightest know the certainty of those things, wherein thou hast been instructed.**

We know that there were hundreds of eyewitnesses who saw the

Lord Jesus alive after He died on the cross. It is not surprising, therefore, to read that "many have taken in hand to set forth in order a declaration ...." so that there would be many written accounts of the events leading up to and following the Crucifixion; but God chose to preserve only the eyewitness accounts from the hands of Matthew, Mark, Luke, and John. We know that those four eyewitnesses were inspired by God because only their accounts were selected to be included in the Bible.

The opening verses of Luke 1 also give us a hint that we are reading a message from God Himself as we read this Gospel. After all, could Luke really be expected to have "perfect understanding of all things" from the very first? God's own words tell us that some things were sealed up until the time of the end, so that Luke could not have understood them. Only God Himself has perfect understanding.

Our Reaction to God's Words

When someone hears God's words, that person may react in many different ways. The Bible gives us two interesting examples of this. Both of these are men who were kings.

Jehoiakim was a king of Judah during one of the times God sent the prophet Jeremiah to warn Judah. Jeremiah's secretary, a man named Baruch, had come to the princes of Judah. He brought with him the scroll containing the words God had instructed Jeremiah to write. The princes brought word of the scroll to the king, and left the scroll in the keeping of a scribe named Elishama. In Jeremiah 36:21-24, we read of king Jehoiakim's arrogance as he heard the words of the scroll being read to him by a man named Jehudi, who brought the scroll to the king:

> **So the king sent Jehudi to fetch the roll: and he took it out of Elishama the scribe's chamber. And Jehudi read it in the ears of the king, and in the ears of all the princes which stood beside the king. Now the king sat in the winterhouse in the ninth month: and *there was a fire* on the hearth burning before him. And it came to pass,**

> *that* when Jehudi had read three or four leaves, he cut it with the penknife, and cast *it* into the fire that *was* on the hearth, until all the roll was consumed in the fire that *was* on the hearth. Yet they were not afraid, nor rent their garments, *neither* the king, nor any of his servants that heard all these words.

Later on in this chapter of Jeremiah, we find that the scroll that king Jehoiakim burned was replaced; it was rewritten and expanded. In Jeremiah 36:32, we read:

> **Then took Jeremiah another roll, and gave it to Baruch the scribe, the son of Neriah; who wrote therein from the mouth of Jeremiah all the words of the book which Jehoiakim king of Judah had burned in the fire: and there were added besides unto them many like words.**

It is this rewritten scroll that has come down to us today in the Book of Jeremiah. There is another interesting point about the account of king Jehoiakim burning the scroll. According to the account, the king destroyed the scroll in portions. A scroll was one very long sheet of material that was unrolled as it was read. The Biblical account tells us that after "three or four leaves" were read to him, the king cut the section out with a penknife and threw it into the fire. These leaves had nothing to do with the sheets of a modern book. So far as we can tell, books were not made like that until about the third century A.D. - some 800 years after king Jehoiakim burned the scroll. In fact, the word translated as "leaves" of the scroll actually means "doors" in Hebrew.

Hundreds of years after the time of Jeremiah, the Lord Jesus told us that He is the door, as we read in John 10:7-9:

> **Then said Jesus unto them again, Verily, verily, I say unto you, I am the door of the sheep. All that ever came before me are thieves and robbers: but the sheep did not hear them. I am the door: by me if any man enter in, he shall be saved, and shall go in and out, and find pasture.**

Just as the Lord Jesus is identified with the door, so is He identified with the word of God. In John 1:1, He is plainly called the Word:

> **In the beginning was the Word, and the Word was with God, and the Word was God.**

King Jehoiakim lived and died hundreds of years before the Lord Jesus' time on earth; but when this haughty king destroyed those "doors" of the scroll - as he destroyed the scroll piece by piece and rejected the word of God - it was almost as if he were among the Pharisees who rejected and ultimately crucified the Lord.

King Jehoiakim's example stands in stark contrast to that of another one of Judah's kings - king Josiah. The Book of 2 Kings tells us about him. He is considered to be one of Judah's great kings. Josiah had a long reign, which ended not long before Jehoiakim became king.

In his reign, king Josiah undertook to repair the temple. This work must have been extensive, for in 2 Kings 22:6 we read about the type of workers to whom silver was to be delivered for the work:

> **Unto carpenters, and builders, and masons, and to buy timber and hewn stone to repair the house.**

During the course of the work, the "book of the law" was discovered; this was the Bible as it existed at that time when it had not yet become the complete book we have today. In 2 Kings 22:8-11, we read what happened:

> **And Hilkiah the high priest said unto Shaphan the scribe, I have found the book of the law in the house of the LORD. And Hilkiah gave the book to Shaphan, and he read it. And Shaphan the scribe came to the king, and brought the king word again, and said, Thy servants have gathered the money that was found in the house, and have delivered it into the hand of them that do the work, that have the oversight of the house of the LORD. And Shaphan the scribe shewed the king, saying, Hilkiah the priest hath delivered me a book. And Shaphan read it before the king. And it came to**

pass, when the king had heard the words of the book of the law, that he rent his clothes.

Based on Josiah's reaction to the reading, we may say that Israel had not done a very good job of studying the law of God for many years. When God first gave Israel the book of the law, while they were still in the wilderness after leaving Egypt, God commanded them to know His law. In Deuteronomy 11:18-20, we read God's instructions to Israel:

Therefore shall ye lay up these my words in your heart and in your soul, and bind them for a sign upon your hand, that they may be as frontlets between your eyes. And ye shall teach them your children, speaking of them when thou sittest in thine house, and when thou walkest by the way, when thou liest down, and when thou risest up. And thou shalt write them upon the door posts of thine house, and upon thy gates:

God had commanded His people to know His words. It was to be their preoccupation. Yet in the passage about Josiah hearing the book read to him, we learn that the book of the law had been discovered - meaning that it had been misplaced. We would certainly expect the high priest and a scribe to know where it was kept at all times, but they didn't. Perhaps king Josiah had been exposed to God's word when it was briefly quoted on some occasions prior to the book's discovery; but now, as Shaphan read the book to him, Josiah really *heard* and understood. He must have been terrified for himself and for his people.

We know that Josiah's was one of God's elect, for in 2 Kings 23:25 we read this about him:

And like unto him was there no king before him, that turned to the LORD with all his heart, and with all his soul, and with all his might, according to all the law of Moses; neither after him arose there *any* like him.

As we continue reading 2 Kings, we find that Josiah instituted many reforms after hearing the book of the law. He was very zealous for God after that.

Jehoiakim and Josiah illustrate the two extremities of the way people may react to hearing the word of God. One individual may react with utter contempt, while another may be cut to the heart and forever changed.

Only God's Words Can Save Us

God uses His words to save people. This is one step or phase in the plan for the salvation of someone who was chosen by God before the foundation of the world. Only God knows exactly when He did this for Josiah, but there is certainly a strong indication that it happened when Shaphan the scribe read the book of the law to him. Josiah may have been exposed to some scriptures earlier in life, but on this occasion he was deeply affected by them.

When God uses His words to save someone, that person changes. If an individual has become saved, that person no longer wants to sin. That person has undergone a fundamental change if the Holy Spirit has applied the word of God to his heart. This is what happened to king Josiah, and this is what happens to someone today when God saves that person. Sadly, today the Holy Spirit is no longer working in the churches to save anyone; so even when the Bible is read there and God's word is preached, there is no possibility that anyone in the congregation can be saved.

Only God's words can save a person. It is true that the writings of man - including writings considered holy among the world's non-Bible based religions - may help improve someone's life greatly. After all, we are each of us a product of our genetic inheritance, our experiences, and the way we exercise our free will. After reading an inspirational story, a self-help book, or one of the world's holy books other than the Bible, a person may turn his life around by an act of his will. Such a book may help someone overcome addictive behavior, improve the way that person treats everyone else, or instill within that person a drive to accomplish great things. Of course, the Bible can do all these things in addition to something that no other book can do: it can save us from the wrath of God. In Isaiah 55:11, God tells us something about His word:

> **So shall my word be that goeth forth out of my mouth:it shall not return unto me void, but it shall accomplish that which I please, and it shall prosper *in the thing* whereto I sent it.**

In this scripture, it almost sounds like God's word is a living thing - as if God sends it out on a mission, and it does the job that God wants it to do. This is also the idea we get when we read what Hebrews 4:12 says about God's word:

> **For the word of God *is* quick, and powerful, and sharper than any two-edged sword, piercing even to the dividing asunder of soul and spirit, and of the joints and marrow, and *is* a discerner of the thoughts and intents of the heart.**

In this scripture, the word "quick" means "living." The Lord Jesus said something about the word of God that helps us understand. In John 6:63 we read:

> **It is the spirit that quickeneth; the flesh profiteth nothing: the words that I speak unto you, *they* are spirit, and *they* are life.**

We need to realize something important. Even though only God can provide understanding of His word, God can save anyone at all. We are not saved because of what we know. God can save an infant, or a person who is deaf and blind, or a person who is not able to think clearly; if the Holy Spirit is present, that person can be saved. We know that the environment in which God saves is the Bible; so as long as a person is in some way exposed to God's words, he or she may be saved.

When the Lord Jesus walked the earth, He spoke words that we may now read in the Bible. Those words of the Lord, and all the other words God caused to be recorded in the Bible - those words and only those words have the power to save someone from God's wrath.

God's Words Condemn Us

It is true that God uses His words, the Bible, to save. However, we are all condemned by God's words because those words - which constitute God's law - require the death penalty for our sins. In Romans 3:23, we read:

For all have sinned, and come short of the glory of God;

Unless God saves us, we remain under the penalty of the law and are subject to God's wrath.

In one of His discourses, the Lord Jesus told His disciples something about the future extent of the Gospel's reach. In Matthew 24:14, we read:

And this gospel of the kingdom shall be preached in all the world for a witness unto all nations; and then shall the end come.

This prophecy is being fulfilled today as never before, as the word of God goes out into all the world. However, many of those who hear the Gospel - God's words - will not be saved. For the unsaved, the Gospel will be a witness against them to condemn them.

It will not only be those who treat the Gospel with contempt - as did king Jehoiakim - who will be condemned. The condemned will include those who know that the Gospel (the Bible) is the word of God and think that they can be saved by keeping its commandments. In the Old Testament, we are given the children of Israel as an example of this kind of thinking. They never did understand that it's impossible for any one to keep the law perfectly. In James 2:10, we read:

For whosoever shall keep the whole law, and yet offend in one *point*, he is guilty of all.

Once an individual has broken the law, the penalty is there; no number or type of good works, not even a lifetime of service to God, can get can him out from under the penalty. Unless you have always kept the whole law perfectly, then you have incurred the penalty of the law and are in need of being saved. The children of Israel thought they could satisfy God by meticulously keeping the commandments. They never understood God's word and their sins, even though God sent them many prophets to warn them. Their wrong attitudes continued right up to the time of the Lord Jesus, as we read in Luke 11:42:

> **But woe unto you, Pharisees! for ye tithe mint and rue and all manner of herbs, and pass over judgment and the love of God: these ought ye to have done, and not to leave the other undone.**

Their kind of thinking continues even today among the Orthodox Jews. Instead of realizing their need for the Saviour, they have adopted a "do-it-yourself" plan by following a long list of "do's" and "don'ts." Instead of realizing that the law's purpose is to drive them to beg God for mercy, they have in their pride rejected the only One through whom they could obtain mercy, as we read in Psalm 118:22:

> **The stone *which* the builders refused is become the head *stone* of the corner.**

As the Gospel currently goes out to the world, we find that those in the local congregations are falling into the same trap. Of course, it may not be so obvious because they have not "officially" rejected the Lord Jesus; but they have rejected His saving work by insisting that it is only by some action on their own part that their salvation can be completed. They have misunderstood and rejected the word of God, and so they will be condemned by it.

God's Gift to Mankind

The Bible is one of God's blessings to us all, both the saved and unsaved. Even an unsaved person who acknowledges God may live in hope of being saved. Such a person could understand something of the purpose of life. He could know that mankind is special, made in the image and likeness of God, and not just another class of primate who is further along the evolutionary scale and whose life is strictly a result of a meaningless, chance occurrence.

Living in accordance with God's word is to our own benefit, to our family's benefit and to the benefit of our world. Just think of how much better the world would be if, for example, everyone decided to follow God's command against stealing. Even with all the other sinful behavior we see all over the world, if everyone observed that one commandment life would be greatly improved for us all.

God Made It Difficult

God could have written the Bible in such a way that even a child in grammar school could understand it, but that's not what He did. In fact, God intentionally made it difficult to understand His word. For example, in 1 Timothy 3:15, we read something from one of the epistles that the apostle Paul sent to his helper, Timothy:

> **But if I tarry long, that thou mayest know how thou oughtest to behave thyself in the house of God, which is the church of the living God, the pillar and ground of the truth.**

Someone may use this verse to claim that the church (meaning his particular denomination with all of its local congregations) is the ultimate authority: "the pillar and ground of the truth." "After all," he may say, "the scripture clearly says that it is the church of the living God that is "the pillar and ground of the truth." This verse, however, can also be understood to mean that it is the living God who is the pillar and ground of the truth. From everything else we read in the Bible, we learn that this verse cannot mean that the local congregations have this trait. Only God can be the pillar and ground of truth; yet it is easy to take this verse to mean something else, if that is your inclination.

When we read in the Bible about God's elect, we learn many details about them, both good and bad. Someone may twist these things to mean something entirely different than what the Bible is teaching. For instance, when Abraham and his wife were going to enter Egypt, he made a decision that we read about in Genesis 12:11-13:

> And it came to pass, when he was come near to enter into Egypt, that he said unto Sarai his wife, Behold now, I know that thou *art* a fair woman to look upon: Therefore it shall come to pass, when the Egyptians shall see thee, that they shall say, This *is* his wife: and they will kill me, but they will save thee alive. Say, I pray thee, thou *art* my sister: that it may be well with me for thy sake; and my soul shall live because of thee.

There is no question that Abraham was one of God's children. Knowing this about him, and reading this account of his lie, someone may conclude that it is permissible to lie under certain circumstances. Of course, this is certainly not what God is teaching us. Even Abraham was a sinner in need of salvation. When we read of Abraham's sin, or the sins of anyone else God has saved according to the Bible, we need to realize that we are all sinners who need forgiveness for our sins.

In Matthew 13:10-11, we read a surprising statement made by the Lord Jesus:

> And the disciples came, and said unto him, Why speakest thou unto them in parables? He answered and said unto them, Because it is given unto you to know the mysteries of the kingdom of heaven, but to them it is not given.

Do these words suggest to you that God is desperately trying to reach everyone in the world with the truth, to make them understand so that they may be saved? That's not how it is. We have to struggle to understand God's words.

God Gives Us Understanding

It isn't just the actual, recorded, spoken words of the Lord Jesus that may be difficult to understand. Whenever we read anything in the Bible, we have to handle it with care. For example, in Acts 2, after he spoke to a multitude on Pentecost, the apostle Peter was asked the question, "What shall we do?" In Acts 2:38, he said:

Then Peter said unto them, Repent, and be baptized every one of you in the name of Jesus Christ for the remission of sins, and ye shall receive the gift of the Holy Ghost.

This sounds like a nice, simple formula for anyone who wants to be saved: repent, and be baptized in water. It is not, however, so simple as that. The only kind of baptism that can save anyone must be done by God. In Luke 3:16, we read something John the Baptist said about this:

John answered, saying unto *them* all, I indeed baptize you with water; but one mightier than I cometh, the latchet of whose shoes I am not worthy to unloose: he shall baptize you with the Holy Ghost and with fire:

Baptism is one of many difficult topics we find in the Bible. We may read something in the Bible and think we have understood it, but did we really? Perhaps we have understood one of that scripture's two or more meanings, without understanding its most important meaning. There are undoubtedly people who have been reading, even studying, the Bible for many years, yet still don't understand the meaning God has placed there. Only God can reveal the Bible's true meaning to us. Some scriptures, such as Proverbs 20:12, seem to have an obvious meaning:

The hearing ear, and the seeing eye, the LORD hath made even both of them.

When we read this scripture, we may think to ourselves, *"Of course God made them; they didn't develop by accident. They are part of God's marvelous creation, and we should praise Him for all these things. And if we have been blessed with good vision and hearing we should thank God for them."* This thinking is certainly appropriate for a Christian. God expects us to be aware that He is the Creator, to praise Him for what He has done and be thankful for all the blessings He has bestowed on us. Our reaction to this scripture, however, should go beyond these thoughts. When we read about the hearing ear and the seeing eye, we should realize that God is telling us that only He can open our ears to hear His words with understanding, and only He can open our eyes to really see the meaning of what we have read when we are reading His words.

God may choose a moment when someone is reading the Bible (or a tract or a book quoting the Bible) to give that person understanding; or perhaps understanding will come upon hearing a voice on the radio reading a scripture while explaining a passage in the Bible. In the Book of Acts, we find an example of how God provided understanding to one of His elect by sending one of His servants to him at just the right time. In Acts 8, a disciple named Philip is told to go south to the way that goes down from Jerusalem to Gaza, which is desert. We read in Acts 8:27-28:

> **And he arose and went: and, behold, a man of Ethiopia, an eunuch of great authority under Candace queen of the Ethiopians, who had the charge of all her treasure, and had come to Jerusalem for to worship, Was returning, and sitting in his chariot read Esaias the prophet.**

As we continue in the chapter, we find that Philip - instructed by the Holy Spirit - runs up to this Ethiopian treasurer and hears him reading; so Philip asks him if he understands what he has read. In Acts 8:31, we read his reply:

> **And he said, How can I, except some man should guide me? And he desired Philip that he would come up and sit with him.**

We learn that the Ethiopian treasurer had been reading about the Lord Jesus in what we know as the Book of Isaiah, but he didn't understand. Philip explained to the Ethiopian the meaning of the passage, for in Acts 8:35, we read:

> **Then Philip opened his mouth, and began at the same scripture, and preached unto him Jesus.**

God used Philip that day to guide the Ethiopian so that his eyes could be opened to the Gospel. God can still use individuals in this way to spread the Gospel, but He is now working through modern methods provided by technology as well.

The Bible is a spiritual book - the only such book in the world. In Romans 7:14, we read:

> For we know that the law is spiritual: but I am carnal, sold under sin.

We can't understand the law, which is the Bible, unless God the Holy Spirit guides us to understand it. Shortly before the Lord Jesus went to the cross, He spoke to His disciples about the Holy Spirit. The Lord promised to send the Holy Spirit to them after He went away. Then, the Lord said something that is very applicable to us today as we seek to understand the Bible. In John 16:13, we read:

> Howbeit when he, the Spirit of truth, is come, he will guide you into all truth: for he shall not speak of himself; but whatsoever he shall hear, *that* shall he speak: and he will shew you things to come.

Surely, the Holy Spirit has been acting on the minds of men in these last days. Our understanding of God's word today is greater than at any other time in history.

The Gospel Is Throughout

Once God has opened our eyes, we may begin to find spiritual truth in many scriptures that don't apparently have any meaning beyond the obvious. Perhaps this is what happened to the Jews of Berea after the apostle Paul went there with Silas. In Acts 17:11, we read this about the Bereans:

> These were more noble than those in Thessalonica, in that they received the word with all readiness of mind, and searched the scriptures daily, whether those things were so.

Some scriptures, for instance, appear to be only simple narrative; yet they illustrate some deep truth that we find elsewhere in the Bible. For instance, the scriptures of John 6:1-3 don't appear to be saying anything spiritual, as we see:

After these things Jesus went over the sea of Galilee, which is *the sea* of Tiberias. And a great multitude followed him, because they saw his miracles which he did on them that were diseased. And Jesus went up into a mountain, and there he sat with his disciples.

Yet we may find in these scriptures much more than first meets the eye. In the words "After these things," God appears to be referring to the previous chapter. There, we read that the Lord told the Jews that He came from the Father, and that the Father had given Him works to finish.

When we read about the Lord crossing over the sea, we need to realize that God uses the sea as a picture of hell. The Lord Jesus had to go through the "sea" of hell before the foundation of the world; that is, He had to endure the wrath of God in order to save the elect.

Then we read that a great multitude followed Him. Elsewhere in the Bible, we read that God has promised to save "a great multitude, which no man could number." Mention of the multitude who followed the Lord Jesus that day suggests the great multitude who will be saved around the time of the great tribulation, during the last days.

In order to save an individual to live for eternity, God has to give that person a new resurrected soul. This is certainly a great miracle. The miracles that the Lord Jesus did on them that were diseased, as we read in those opening verses from John 6, suggest this salvation. After all, the sin which has made us spiritually dead is like a loathsome disease in God's sight. Finally, we read that the Lord Jesus went up into a mountain and sat with His disciples. Isn't this a picture of the Kingdom of God after the end of the world? God's elect will be there with Him for all time.

Anyone whom God permits to find hidden truth in this way should certainly consider himself to be blessed. Since these truths are consistent throughout the Bible, when we find hidden truth we know that the writing of the entire Bible was guided by a single mind - the mind of God. Only God could have concealed truth in this manner; and so we have further proof that only God could have written the Bible.

How Should We Read the Bible?

God wrote the Bible in a way that is different than anything that anyone else has ever written. He wrote to reveal, but also to conceal. He wrote to save, but also to condemn. He wrote to comfort, but also to warn. Because the Bible is so different than any other book, one might reasonably ask if it must be read differently. The Bible can't be read as any ordinary book. God gives us some rules about how we are to read it. For one thing, we need to understand that every word of the Bible, in the original language in which it was written, is the word of God. In 2 Timothy 3:16, we read:

All scripture *is* given by inspiration of God, and *is* profitable for doctrine, for reproof, for correction, for instruction in righteousness:

The original Greek, translated in this scripture as "given by inspiration of God," carries the idea that scripture is "God breathed." It's as if each word came from the mouth of God. God also took care to be sure that the Bible was preserved in tact - every word of it - down through the ages. He raised up a class of Jewish scribes who faithfully and accurately copied out the exact words, checking and rechecking so that there would be no error.

If we come across a scripture that appears to contradict something else we have read in the Bible, we need to dig (and pray!) a bit deeper for understanding. A very valuable tool for Bible study with the King James Version of the Bible is Strong's Exhaustive Concordance of the Bible (Large Print Edition). The concordance is an amazing book that allows you to find every scripture in the Bible that uses a given word. By examining the way a word is used elsewhere in the Bible, you may be better able to understand the sense in which it is used in a scripture you are studying. This approach to the Bible works because the Bible is a cohesive whole, all coming from a single source. A concordance will also help you determine the meaning or meanings of the

Hebrew or Greek word that has been translated into the word you are investigating.

An illustration from science may prove useful here. Sometimes, a scientist who is collecting data from an experiment will get a data point that is totally out of line with all the others. He may just "throw it out," thinking that something must have gone wrong with that measurement and that it wasn't any good. We must never do that with anything we read in the Bible. Someone may dismiss a scripture as containing a scribal error because it doesn't seem to fit with the others he has studied; but that scripture, like everything else in the Bible, is "God breathed." It may be that the very scripture we are struggling to understand is the key to the truth we are seeking.

In Isaiah 28:10, we read an important rule of Bible study:

For precept *must be* upon precept, precept upon precept; line upon line, line upon line; here a little, and there a little:

In this way, we might think of the Bible as being like a giant jig saw puzzle. When we try to fit a scripture into the framework of our thinking about a certain doctrine and eventually learn how it fits, we begin to see the "big picture" that the Bible is teaching. In 1 Corinthians 2:12-13, we are also given a rule about Bible study:

Now we have received, not the spirit of the world, but the spirit which is of God; that we might know the things that are freely given to us of God. Which things also we speak, not in the words which man's wisdom teacheth, but which the Holy Ghost teacheth; comparing spiritual things with spiritual.

We are to compare spiritual things with spiritual things; that is, we are to compare one scripture with another. Most of the time, in order to follow this rule when we are considering a scripture, we will want to compare it with other scriptures that use one or more of the same words. The concordance is invaluable in this task.

When we find each way in which a particular Hebrew or Greek word is used in the Bible, we are letting the Bible act as its own dictionary. This is as it should be, and it follows logically from the rule about "comparing spiritual things with spiritual."

For example, suppose you find in your King James Bible a scripture having a word that strikes you as being interesting or strange. With the concordance, you can find the Hebrew or Greek word that was translated into that English word. When you see the associations of that Hebrew or Greek word in its various contexts - that is, when you see what other words are used with that Hebrew or Greek word in the various scriptures - you may begin to understand its meaning.

In many cases, the translators have rendered several Hebrew or Greek words into the same English word. We must therefore be careful not to confuse that Hebrew or Greek word with another Hebrew or Greek word that has also been translated into the English word we are checking. In order for the Bible to serve as its own dictionary, we need to find the Hebrew or Greek word that was translated into the English word we are checking, and then check the other uses of that same Hebrew or Greek word. Then, possibly, after a while we may begin to understand how the scripture in question can be reconciled with another scripture that seems to be stating something contrary.

Translation Errors

When we say that the Bible is the word of God, we mean that the original Hebrew and Greek words are God-breathed. God did not promise to be with any translators, and so we should not be shocked to learn that our Bibles contain some translation errors. Some of these errors can be discovered by using a concordance; others can't. Consequently, there may be times when we are struggling to understand a scripture and it will not be possible to arrive at a correct understanding of it without going back to the original Hebrew or Greek - unless we have previously been made aware that there is a problem with the way that particular verse has been translated. For very many readers of the Bible, it will not be practical or even possible to read a scripture in its original language. However, it is worth keeping in mind that the various translations do contain some errors. That being said, we can - in most cases - trust the King James Version to provide an accurate translation. For example, in Isaiah 9:3, we read:

Thou hast multiplied the nation, *and* not increased the joy: they joy before thee according to the joy in harvest, *and* as *men* rejoice when they divide the spoil.

If you were to read this same scripture in the New International Version (NIV) of the Bible, you would read the following:

You have enlarged the nation and increased their joy; they rejoice before you as people rejoice at the harvest, as men rejoice when dividing the plunder.

Immediately we see a glaring difference between the two translations. According to the King James Version, the joy has not increased; according to the NIV, it has. Obviously, one of the two translations is in error. The King James translators got it right. Apparently, the scripture didn't make sense to the NIV translators - so they changed it to what they thought it should say! (You may be able to verify this for yourself by using an interlinear Bible. You can do this from your computer if you have Internet access. The interlinear Bible shows you the scripture as it appears in English in the King James Version, and below it you will see the scripture in Hebrew. By looking at several scriptures containing the word "not," you should be able to identify the Hebrew word for it. It contains two characters: from left to right, one vaguely resembling an "x" and the other a "7" with a small vertical addition at the top where the character begins. Note that Hebrew is written from right to left.)

We could never arrive at a correct understanding of this scripture by using the NIV translation of it. (This scripture seems to be anticipating the church age, when God's word would be spread around the world as Bibles were multiplied. Despite this, relatively few people were saved during the church age, and so the joy did not increase.) The NIV translators responsible for this error did not approach the Bible as they should have. They seemed to forget that it is the word of God.

Chapter 2 God's Book

The Bible warns anyone who might be tempted to change its words. In Revelation 22:18-19, we read:

For I testify unto every man that heareth the words of the prophecy of this book, If any man shall add unto these things, God shall add unto him the plagues that are written in this book: And if any man shall take away from the words of the book of this prophecy, God shall take away his part out of the book of life, and out of the holy city, and *from* the things which are written in this book.

If we think about these scriptures, we should realize that they apply to the entire Bible - not just to the Book of Revelation. To understand why this is so, we need to realize that it was God's intention, even before the foundation of the world, to give mankind a book - the Bible. God didn't give it to us all at once, though; He gave it to us in installments. Some of Charles Dickens' novels were published in installments, so the idea of publishing a book in this manner would not have been at all unusual to his readers. When it came to the Bible, however, the "installments" were spread over many centuries. God gave the first few of them to Moses "for publication." He gave the last one to the apostle John - that was the Book of Revelation. In between there were 60 more installments given to many different men.

Once the Book of Revelation got out to the world, mankind had the whole book. It was in that final installment that God warned us neither to add nor to take away anything from His book. God's message to mankind was now complete. There would be no further communication from Him in a heavenly language - and this is what the phenomenon of speaking in tongues was all about when it occurred early in the church age, before the Bible was complete; there would be no further revelation from Him by dreams or visions; He would give no more signs or wonders. God has alerted us so that if any of these things did occur, we would know that Satan is responsible.

The insertion or deletion of words from any part of the Bible is not the only way to add to or take away from the things in the Bible. If we are relying on someone claiming he has received dreams, or visions or some other message from God, then we are violating this part of God's law. The Bible alone, and in its entirety, must be our authority. Whenever we read the Bible or speak about it or attempt to teach it, we need to keep this warning from Revelation 22 in mind.

The King James version of the Bible, although superior to other translations, is not entirely without translation errors. For example, in Matthew 28:1, we read about the first visitors to the empty tomb on the day the Lord Jesus arose:

> **In the end of the sabbath, as it began to dawn toward the first *day* of the week, came Mary Magdalene and the other Mary to see the sepulchre.**

The word "sabbath" here should have been translated as "sabbaths." The phrase "first day of the week" is also a problem: it should have been translated as "first of the sabbaths." (The word "day" appears in italics to denote that it was not in the original language - it was added by the translators.) Putting it all together, we get the following translation:

> In the end of the Sabbaths, as it began to dawn toward the first of the Sabbaths, came Mary Magdalene and the other Mary to see the sepulchre.

This is how that verse should have been translated. If you have access to an interlinear translation, you can see this for yourself. You may be able to find an easy-to-use interlinear translation on the Internet. To detect the problem, you will have to download the Greek font used in the original language. After you can pick out the word for sabbath near the beginning of the verse, you will see that it is repeated where the translators used the phrase "first day of the week." Take a look at the Greek word translated as "sabbath." It is in the plural in both places. You can detect the plural form by looking at the next to the last letter of the word. For comparison, check out the original language for Mark 9:41 and Mark 7:4, looking at the words for "cup" and "cups." In Mark 9:41, the word is cup - singular; but in Mark 7:4, it is plural - cups. Notice where the difference is. It is the same for the plural of the word "sabbath" in Matthew 28:1.

Why should we be so concerned about this error in translation? As it happens, Matthew 28:1 is a key verse in showing us that the era of the seventh day sabbath had passed, and that a new era had begun: the era of the Sunday sabbath.

Matthew 28:1 is just one of the scriptures that King James' translators didn't know how to handle. There are others. That being said, we can still be confident that the King James version is overall a fine translation. In fact, it is not only superior to the NIV in the way it has rendered many scriptures, it is also superior in completeness.

There are a number of scriptures you will find in the King James version, that you will not be able to find in the text of the NIV, although they may appear in footnotes. The following is a list of 16 scriptures that aren't in the text of the NIV:

Matthew 17:21	Mark 9:44	Luke 17:36	Acts 15:34
Matthew 18:11	Mark 9:46	Luke 23:17	Acts 24:7
Matthew 23:14	Mark 11:26	John 5:4	Acts 28:29
Mark 7:16	Mark 15:28	Acts 8:37	Romans 16:24

For non-native speakers of English, the King James version takes some "getting used to." God can, of course, save someone who is using the NIV as readily as He can save someone who is using the King James version of the Bible; but for the reasons presented here, the King James version is preferable to the NIV.

A final point should be made about the various Bible translations available today. Some people favor a paraphrase translation. These translations are certainly easy to read and seem to make the Bible easy to understand; but they are not the word of God. Earlier, we saw that God has concealed spiritual truth throughout the Bible. Even in Old Testament events, persons, places or things, we may see a picture of the saving work of the

Lord Jesus. In all likelihood, you would not find that truth in a paraphrase translation. The very rewording that makes a paraphrase translation "flow" so nicely obliterates truth.

In the King James version, sentences may at times seem awkward and peculiar; but it this awkwardness and peculiarity that often reflect the very words God chose to use when He wrote the Bible, and those are words that contain spiritual truth. For example, if we go to Ruth 3:1 as it appears in the Good News Translation of the Bible (Second Edition, 1992, American Bible Society), we read:

> Some time later Naomi said to Ruth, "I must find a husband for you, so that you will have a home of your own.

Compare this with Ruth 3:1 from the King James Version:

> **Then Naomi her mother in law said unto her, My daughter, shall I not seek rest for thee, that it may be well with thee?**

The rendering from the King James version seems very awkward compared to the Good News Translation of this scripture. If we were considering an ancient story written by men, the Good News Translation paraphrase would be just fine. It tells us about Naomi's intentions, as we see when we continue to read Ruth chapter 3. There we learn that Naomi has developed a plan by which Ruth can demonstrate her willingness to marry a man named Boaz. Nevertheless, the paraphrase translation doesn't accurately reflect what Namoi said. Naomi's words seem peculiar to us, and indeed they may have been strange even for someone living at that time in those circumstances. For Naomi to say that she seeks "rest" for Ruth doesn't really sound right to our ears; but if we remember other scriptures that tell us about rest, we may see a valuable spiritual lesson. God tells us about entering into His rest in Hebrews 4:10, where we read:

> **For he that is entered into his rest, he also hath ceased from his own works, as God *did* from his.**

When God saved Ruth, she entered into His rest - as does anyone whom God has saved. We can see this by comparing other scriptures with Ruth 3:1 as we are reading the King James version of the Bible; but this spiritual lesson would be lost to us if we were reading the paraphrase translation.

Progressive Revelation

In Acts 11, we read a very interesting account of the apostle Peter explaining to the disciples at Jerusalem something that had happened while he was in the city of Joppa. Peter told them of a vision he had seen. In the vision, various animals came down from heaven on what seemed like a great sheet. As this was happening, Peter heard a voice saying "Arise, Peter; slay and eat." The animals Peter saw were of kinds that religiously observant Jews consider to be unclean - they would never eat them; so Peter objected to what he had heard. The account continues in Acts 11:9-10:

> **But the voice answered me again from heaven, What God hath cleansed, *that* call not thou common. And this was done three times: and all were drawn up again into heaven.**

Shortly after this, Peter learned the meaning of the vision God had shown him, as Peter explained to the disciples. He told them that three men had come from Caesarea to see him. These men had been sent by a man named Cornelius, who also had seen a vision from God. In his vision, Cornelius saw an angel who told him to send men to Joppa to call for Peter.

Peter and six others accompanied the men from Joppa and came to Cornelius' house. Cornelius explained to Peter why he had sent for him; Peter told all this to the disciples, as we read in Acts 11:13-18:

> **And he shewed us how he had seen an angel in his house, which stood and said unto him, Send men to Joppa, and call for Simon, whose surname is Peter; Who shall tell thee words, whereby thou and all thy house shall be saved. And as I began to speak, the Holy Ghost fell on them, as on us at the beginning. Then**

> remembered I the word of the Lord, how that he said, John indeed baptized with water; but ye shall be baptized with the Holy Ghost. Forasmuch then as God gave them the like gift as *he did* unto us, who believed on the Lord Jesus Christ; what was I, that I could withstand God? When they heard these things, they held their peace, and glorified God, saying, Then hath God also to the Gentiles granted repentance unto life.

"Then hath God also to the Gentiles granted repentance unto life." With these words, the disciples realized that God wasn't limiting salvation to the Jews. When we read the Old Testament, we see indications even hundreds of years before Peter's time that this was God's intention; but the early New Testament church didn't understand it because God still hadn't revealed it to them. God waits until the right time, according to His timetable, to reveal truths hidden in the scriptures.

In these last days of the earth's existence, God has revealed many amazing truths. Perhaps there is more to come, perhaps not. We do know that if there are more truths to be revealed, God will do it in the very near future because we are almost out of time.

The Biblical teaching that God would in time reveal more truth is referred to as progressive revelation. A clear statement of this is found in the Book of Daniel, in which the prophet Daniel is given in a vision many prophecies pertaining to the end-times. In Daniel 12:8-10, we read:

> And I heard, but I understood not: then said I, O my Lord, what *shall be* the end of these *things*? And he said, Go thy way, Daniel: for the words *are* closed up and sealed till the time of the end. Many shall be purified, and made white, and tried; but the wicked shall do wickedly: and none of the wicked shall understand; but the wise shall understand.

Daniel wanted to understand, but he couldn't; so he asked for more information. In response, he was told that the words would be "closed up and sealed till the time of the end," and that the "wise shall understand." Since the day Daniel was told this, God has opened up His word to reveal the actual dates pertaining to end-time judgment and the end of the world; and not only the dates - God has revealed the nature of His judgment as well.

The fact that God has recently revealed His timetable for end-time events and many details about the nature of His coming judgment against mankind is in accord with what we read in the scriptures. In Ecclesiastes 8:5, we read:

> **Whoso keepeth the commandment shall feel no evil thing: and a wise man's heart discerneth both time and judgment.**

We should also be aware of another promise God has made. In Amos 3:7, we read:

> **Surely the Lord GOD will do nothing, but he revealeth his secret unto his servants the prophets.**

Today a prophet is anyone who is one of God's elect. In these end times, God is revealing His secrets by opening up the Bible to our understanding. The words haven't changed, but our understanding of them has been changing by the work of the Holy Spirit. In this way, we have received the incredible knowledge that God will soon destroy the earth.

You may have seen one of those large electronic message boards in a public waiting room, in an airport or in a store. As you watch it, you will see a variety of messages applicable to the location and purpose of the facility. Well, the Bible is sort of like that. In Luke 21:36, we read:

> **Watch ye therefore, and pray always, that ye may be accounted worthy to escape all these things that shall come to pass, and to stand before the Son of man.**

We watch by reading the Bible, because what the Bible can tell us will vary with time. If we are living at a time when God is preparing to do something - and this is certainly the case in our day - then God will reveal it through His word.

Chapter 3

God's Law

God created a universe that is governed by laws. For example, the gravitational force between two objects is inversely proportional to the square of the distance between them; and the volume a gas will occupy is directly proportional to its temperature. There are undoubtedly many thousands of such relationships. When such a relationship is discovered, it can be formulated into a mathematical equation. That's why mathematics is such a useful tool for scientists and engineers - because the universe follows laws. Well, there are other types of laws besides these. There are spiritual laws, and the Bible tells us about them.

Many non-Christians, when they are learning about Christianity for the first time, have undoubtedly wondered why the Lord Jesus had to die to save anyone from their sins. After all, if He's God, He can do anything, right? Well, God's law *is* the reason why the Lord suffered and died. Just as man is subject to God's laws, God has made Himself subject to His own spiritual laws. This law demands a penalty for sin, and God paid that penalty on behalf of those He chose to save.

The Ten Commandments

Upon hearing any mention of "God's laws," many people will immediately think of the Ten Commandments. Actually, the entire Bible is the law of God; but the Ten Commandments have achieved special prominence. These commandments are listed twice in the Bible - in Exodus chapter 20, and again in Deuteronomy chapter 5. The following is the list from Exodus 20:2-17:

I *am* the LORD thy God, which have brought thee out of the land of Egypt, out of the house of bondage. Thou shalt have no other gods before me.

Thou shalt not make unto thee any graven image, or any likeness *of any thing* that *is* in heaven above, or that *is* in the earth beneath, or that *is* in the water under the earth: Thou shalt not bow down thyself to them, nor serve them: for I the LORD thy God *am* a jealous God, visiting the iniquity of the fathers upon the children unto the third and fourth *generation* of them that hate me; And shewing mercy unto thousands of them that love me, and keep my commandments.

Thou shalt not take the name of the LORD thy God in vain; for the LORD will not hold him guiltless that taketh his name in vain.

Remember the sabbath day, to keep it holy. Six days shalt thou labour, and do all thy work: But the seventh day *is* the sabbath of the LORD thy God: *in it* thou shalt not do any work, thou, nor thy son, nor thy daughter, thy manservant, nor thy maidservant, nor thy cattle, nor thy stranger that *is* within thy gates: For *in* six days the LORD made heaven and earth, the sea, and all that in them *is*, and rested the seventh day: wherefore the LORD blessed the sabbath day, and hallowed it.

Honour thy father and thy mother: that thy days may be long upon the land which the LORD thy God giveth thee.

Thou shalt not kill.

Thou shalt not commit adultery.

Thou shalt not steal.

> **Thou shalt not bear false witness against thy neighbour.**
>
> **Thou shalt not covet thy neighbour's house, thou shalt not covet thy neighbour's wife, nor his manservant, nor his maidservant, nor his ox, nor his ass, nor any thing that *is* thy neighbour's.**

The Ten Commandments were written by God Himself in two tables of stone when Moses was on mount Sinai. We may see these commandments inscribed on many monuments in state capitols, and framed on county courthouse walls. There is also a very famous and commercially successful, though not so Biblically accurate, movie by that name. Perhaps many people with no religious background whatsoever have seen, or at least heard of, this film and in that way know of the Ten Commandments.

The Whole Bible is a Book of Laws

A great many people associate the Ten Commandments - and the Ten Commandments exclusively - with God's laws. This is because they fail to realize that God has placed many other laws throughout the Bible. In fact, God is a God of law, and the whole Bible is a book of laws. For example, we are told to live by every word that proceeds from the mouth of God, as the Lord Jesus said in Matthew 4:4:

> **But he answered and said, It is written, Man shall not live by bread alone, but by every word that proceedeth out of the mouth of God.**

This in itself is a command, and it is a command that is telling us to live by every word of God; but the whole Bible is God's word! Therefore, we are to study God's word - the whole Bible - to know what God has commanded us to do. This is where we will find truth, and it is the only source we can trust.

There are laws throughout the Bible, so we cannot assume that we are "right with God" because of our attempts to keep the Ten Commandments - no matter how successful we think we may be in keeping them; besides, we will see that there is much more to these commandments than we can know just by reading them in the Old Testament.

When the Lord Jesus was asked which commandment is first of all, He replied as recorded in Mark 12:29-31:

> **And Jesus answered him, The first of all the commandments** *is*, **Hear, O Israel; The Lord our God is one Lord: And thou shalt love the Lord thy God with all thy heart, and with all thy soul, and with all thy mind, and with all thy strength: this** *is* **the first commandment. And the second** *is* **like,** *namely* **this, Thou shalt love thy neighbour as thyself. There is none other commandment greater than these.**

In the Old Testament, we read laws that are very close to these two stated by the Lord Jesus, although they are in different books of the Bible. In Deuteronomy 6:4-5, we read:

> **Hear, O Israel: The LORD our God** *is* **one LORD: And thou shalt love the LORD thy God with all thine heart, and with all thy soul, and with all thy might.**

In Leviticus 19:18, we read something similar to the second great commandment as it was stated by the Lord Jesus:

> **Thou shalt not avenge, nor bear any grudge against the children of thy people, but thou shalt love thy neighbour as thyself: I** *am* **the LORD.**

Those who were there that day to witness the Lord Jesus respond to the question about the greatest commandment should have been absolutely astounded. How could the son of a carpenter be so familiar with the scriptures as to be able to quote them like that, and to have such a profound understanding that He could pull these two concepts from two areas of the scriptures and unite them perfectly? This in itself should have been a sign to them that the person they were hearing was no ordinary man.

Insight into the Ten Commandments

When we keep in mind the Lord's statement of the two greatest commandments, we may notice that the Ten Commandments consist of two major parts. The first part - the first four commandments - deal with man's relationship to God. The second part - the last six commandments - deal with man's relationship with his fellow man. The two great commandments that the Lord Jesus stated in Mark 12:29-31 include, but are not limited to, the Ten Commandments.

We gain further insight into the Ten Commandments when we read some other statements that the Lord Jesus made. Regarding the command against murder, the Lord Jesus stated in Matthew 5:21-22:

> **Ye have heard that it was said by them of old time, Thou shalt not kill; and whosoever shall kill shall be in danger of the judgment: But I say unto you, That whosoever is angry with his brother without a cause shall be in danger of the judgment: and whosoever shall say to his brother, Raca, shall be in danger of the council: but whosoever shall say, Thou fool, shall be in danger of hell fire.**

And regarding the command against adultery, He stated in Matthew 5:27-29:

> **Ye have heard that it was said by them of old time, Thou shalt not commit adultery: But I say unto you, That whosoever looketh on a woman to lust after her hath committed adultery with her already in his heart. And if thy right eye offend thee, pluck it out, and cast *it* from thee: for it is profitable for thee that one of thy members should perish, and not *that* thy whole body should be cast into hell.**

Even your thoughts can get you into trouble with God according to this scripture. This scripture, therefore, reinforces the Bible's teaching that any sexual activity outside of marriage is sinful.

Whether these two statements of the Lord's are clarifications of the sixth and seventh commandments or amplifications of them - actual changes in the law - need not concern us. The Lord Jesus, as God, could write new laws if He wanted to do so. In fact, as we shall see, there *are* new laws in the New Testament. In any event, the Lord's statements about the commands against adultery and murder show us the difficulty of complying with these laws. You say you've never committed adultery or murdered anyone? But have you ever kept lustful thoughts or hatred for someone in your mind? It's one thing to control your actions, but to control your thoughts is so much more difficult. And this is what God is telling us we must do in order to comply with His laws.

There is a scripture showing a relationship between two of the Ten Commandments. In Colossians 3:5, we read:

Mortify therefore your members which are upon the earth; fornication, uncleanness, inordinate affection, evil concupiscence, and covetousness, which is idolatry:

In this scripture, we see that God considers covetousness to be idolatry. It is the tenth commandment that prohibits covetousness; but if we break that commandment, we are putting something - whatever it is that we covet - ahead of God. When we do that, we are also breaking the first commandment. To covet is equivalent to making a god of something and putting it ahead of God Himself. Covetousness also seems to be a violation of the second commandment, which prohibits the making of a graven image, and bowing down to an image to worship it. If we covet something, aren't we in a sense carrying it's image around with us in our minds, and worshiping it? (A further word regarding the second commandment: any artistic work which is a depiction of the Lord Jesus, or the Virgin Mary, or any of the apostles, or any person in connection with religious worship, is a violation of this commandment.)

From this comparison of covetousness with idolatry, we should see that we may be breaking two or more commandments whenever we sin. For example, someone who is planning to steal something is coveting it and therefore breaking the tenth commandment. If, while planning the theft, he has spoken deceitfully to the person from whom he is planning to steal the object, wouldn't he in a sense be a false witness (a violation of the ninth commandment, which forbids lying)? By actually stealing the object, he is breaking the eighth commandment. And wouldn't the act of stealing be dishonoring his parents - a violation of the fifth commandment? We already know that covetousness is idolatry, so it seems we also have a violation of the first and second commandments in this situation. You may have heard the expression "the judge threw the book at him" in connection with a person who was charged with multiple crimes resulting from a single incident. Well, this is our situation as we stand before God, having broken His law whenever we sin.

We should also consider the third commandment - the command against taking God's name in vain. Sadly, this commandment is broken regularly by so many people in their regular manner of speaking, especially when swearing. Whenever we say the name "Jesus," or "Christ," it should be with the utmost reverence - because He is God. In Philippians 2:9-11, we read:

> **Wherefore God also hath highly exalted him, and given him a name which is above every name: That at the name of Jesus every knee should bow, of *things* in heaven, and *things* in earth, and *things* under the earth; And *that* every tongue should confess that Jesus Christ *is* Lord, to the glory of God the Father.**

Of course, some people may not even realize they are breaking the third commandment when they exclaim "Lord," or "God;" but they are. Whenever these words pass our lips, we should remember that we are speaking of God Himself.

As exclamations, many people use euphemisms for God. There are undoubtedly a great many of these throughout the world's various languages. In English, you may have heard someone exclaim "golly," or "gee," or "gee-whiz," or "gosh," or even "crikey," the last being a common Australian exclamation. The use of any of these, or any like them, is a violation of the third commandment.

We may now start to think that it isn't so easy to keep the Ten Commandments after all. In fact, to keep them the way God commands us to keep them is impossible. Perhaps you think that perfection isn't really required. "After all," you may say, "God knows we're not perfect. He'll forgive me for the small sins I commit." The Bible, however, shows us that God sees sin differently than we do. In fact, *any* sin at all makes us guilty in God's eyes. This is what we read in James 2:10:

> **For whosoever shall keep the whole law, and yet offend in one *point*, he is guilty of all.**

According to this verse, in order for anyone to be justified under the law that person must keep every command in the Bible and do it perfectly. Even if the person doesn't understand something about God's law or isn't aware of it, the responsibility to keep the whole law remains. God is perfect, and He can't accept imperfection.

Purpose of the Law

What then are we to do, knowing that we can't possibly keep all God's laws perfectly? We know that God will punish sin, and a sin occurs whenever one of God's laws is broken, as we read in 1 John 3:4:

> **Whosoever committeth sin transgresseth also the law: for sin is the transgression of the law.**

God's law should cause us to acknowledge that we are sinners in need of salvation, and drive us to beg God in prayer for His mercy while it is still the day of salvation. We know that God is merciful, and so long as a person is alive, there is hope for that person - hope that God will save him or her. After a person has died, however, there is no more hope. Either the person has been saved, or he hasn't.

Two Different Types of Laws and the Fourth Commandment

There remains one of the Ten Commandments that has not yet been covered here in any detail: the fourth commandment. What about this commandment - the command to keep the seventh day Sabbath? This is the command that prohibits any kind of work on Saturday, the seventh day of the week. Before considering it, we need to take a look at an aspect of God's laws.

We may categorize God's laws into two different types: moral laws and ceremonial laws. As time has unfolded, God's ways of dealing with mankind have changed in accordance with His salvation plan, which He developed before creation. These changes included the elimination of Old Testament ceremonial laws. (The moral laws, such as the command against murder, have not changed.)

For example, God required ancient Israel to make animal sacrifices to Him. These were ceremonial laws. Animal sacrifices provided ancient Israel with a picture of the sacrifice that the Lord Jesus made before the foundation of the world, and then demonstrated for us by being crucified. Sadly, Israel never understood the significance of the sacrifices.

The requirements for animal sacrifices appear very extensively and in great detail in the Old Testament. There was no time limitation for these sacrifices. God didn't specify a certain number of years that the sacrifices were to be made. Neither did He specify an event that would mark the end of the sacrifices. When reading these laws, you will see that God commands them to be done without end. Yet when the Lord Jesus came, He fulfilled the requirements of those laws. As He stated in Matthew 5:17:

> **Think not that I am come to destroy the law, or the prophets: I am not come to destroy, but to fulfil.**

Therefore, in the New Testament, there are no more animal sacrifices. God also required ancient Israel to establish a priesthood, beginning with Aaron, and gave them many laws pertaining to it. For example, in Exodus 30:8, we read:

And when Aaron lighteth the lamps at even, he shall burn incense upon it, a perpetual incense before the LORD throughout your generations.

God required a perpetual burning of incense upon the altar He had ancient Israel make. This was one of many things the priests had to do, in addition to the animal sacrifices. Yet, hundreds of years later, God Himself made it impossible for His people to keep the priestly commands begun with Aaron when He caused the temple to be destroyed in 587 B.C. by the Babylonians. By the time the Lord Jesus appeared, there was a new temple, and so these ceremonial laws had resumed. But again the temple was destroyed - this time in 70 A.D. by the Romans - and all these activities came to an end. When we read the New Testament, we learn that the Lord Jesus is our High Priest; and so He is the fulfillment of the ancient priesthood begun with Aaron. All of the ceremonial laws pertaining to the priesthood have been fulfilled in the Lord Jesus.

Now, what about the fourth commandment? Are we still required to keep it? In Exodus 31:13, we read:

Speak thou also unto the children of Israel, saying, Verily my sabbaths ye shall keep: for it *is* a sign between me and you throughout your generations; that *ye* may know that I *am* the LORD that doth sanctify you.

If God sanctifies someone, that person is saved. In John 17:16-17, we read part of a prayer by the Lord Jesus regarding those whom the Father chose to be saved:

They are not of the world, even as I am not of the world. Sanctify them through thy truth: thy word is truth.

The Lord prayed that we be sanctified through God's truth. God's word is truth; but the Bible tells us that the Lord Jesus Himself is the Word of God, as we read in John 1:1:

In the beginning was the Word, and the Word was with God, and the Word was God.

Therefore, we are sanctified by the Lord Jesus. This agrees with

Chapter 3 God's Law

what we have read about the fourth commandment - that it is a sign that God sanctifies us.

Just as the fourth commandment was a command not to do any work on one particular day every week, similarly we are not to try to do any work - or even think that we can do any work - to gain our salvation. It is God who sanctifies those He wants to save; He does all the work. In fact, all the work needed to save any one has already been done by the Lord Jesus. It is His work that sanctifies us. We must never think that any work we do will merit salvation, that we can sanctify ourselves. Nothing we do can save us.

Sadly, the local congregations of today have been deceived by a false gospel and think they can and must do something for salvation. They may claim to realize the Lord Jesus did all the work to save them; but you may soon hear something else as well. For example, they may say that salvation is sort of like the gift of an enormous sum of money, presented to you as a check. You must accept the check in order to benefit by it. God's word tells us that salvation doesn't work that way. When we compare scripture with scripture, we learn that those who receive the Lord - those who "accept the check" - are only those whom God has already chosen. Only they can hear the shepherd's voice. It is only they who are the lost sheep for whom the Lord has come. That is how we must understand John 1:12-13:

> **But as many as received him, to them gave he power to become the sons of God, *even* to them that believe on his name: Which were born, not of blood, nor of the will of the flesh, nor of the will of man, but of God.**

Those who receive Him will become the sons of God, but only those whom God has already decided to save will receive Him. We cannot become God's children by our own decision, by the "will of the flesh," as the scripture says. If you believe it is up to you whether or not to "accept the check," then you believe a false gospel, because anything we do is work on our part - and our work can't save us. We need to remember the way the Lord chose His apostles. For example, in Matthew 4:18-20, we read about Peter and Andrew who were busy at their livelihood when they first saw the Lord:

> **And Jesus, walking by the sea of Galilee, saw two brethren, Simon called Peter, and Andrew his brother, casting a net into the sea: for they were fishers. And he saith unto them, Follow me, and I will make you fishers of men. And they straightway left *their* nets, and followed him.**

There was no interview, request or invitation by the Lord. He spoke, and it happened. We should also remember the Lord's parable about the man who prepared a feast. He sent his servant to invite many, but they all had excuses why they couldn't come. In Luke 14:21-24, we read the rest of the parable:

> **So that servant came, and shewed his lord these things. Then the master of the house being angry said to his servant, Go out quickly into the streets and lanes of the city, and bring in hither the poor, and the maimed, and the halt, and the blind. And the servant said, Lord, it is done as thou hast commanded, and yet there is room. And the lord said unto the servant, Go out into the highways and hedges, and compel *them* to come in, that my house may be filled. For I say unto you, That none of those men which were bidden shall taste of my supper.**

Notice that the first group was *brought* in, and the second group was *compelled* to come in. We need to recognize that it's all up to God, and wait on Him to save us.

The fourth commandment, like the other ceremonial laws, was fulfilled by the Lord Jesus. The ancient Jews were required to strictly observe it; but for us it is a sign that the Lord does all the work to save us. We are no longer required to refrain from work on Saturday. We need to be aware of the meaning of the fourth commandment, but we don't have to keep it.

The Importance of God's Law

Some Christians may believe the law should not concern them and that it isn't in force anymore. They are convinced that they have been saved and are under God's grace - not under the law. Someone who has been saved *is* under God's grace, so that the penalty required by the law does not threaten that person any more. In that sense, a saved person is no longer under the law. This, however, does not mean that we don't have to be concerned about God's law. If someone is truly saved, that person should be more concerned than ever about keeping God's law. In fact, God tells us that if we love Him, we will keep His commandments. In John 14:15, the Lord Jesus said:

If ye love me, keep my commandments.

The commandments are the laws of God - the Bible! God places the utmost importance on His law, even putting it above His own name. Even at great cost to Himself, in His integrity, God won't violate the law. In Psalm 138:2, we read:

I will worship toward thy holy temple, and praise thy name for thy lovingkindness and for thy truth: for thou hast magnified thy word above all thy name.

In many scriptures throughout the Bible, we read of the importance that God places on His glory. For example, in I Chronicles 16:29 we read:

Give unto the LORD the glory *due* unto his name: bring an offering, and come before him: worship the LORD in the beauty of holiness.

God places great importance on His glory - on His name. Yet He puts His word, which is His law, above His name. We may be certain that God will execute the provisions of His law.

The Lord Jesus Fulfilled the Law

The Bible tells us that God chose to have a people for Himself. There was, however, a problem: the law demands that everyone - for we are all sinners - pay the penalty for our sins. In order to meet the law's requirements for those people God chose to save, the Lord Jesus had to incur the wrath of God. This was necessary because, if those people had to pay for their sins by forfeiting their lives forever, God couldn't accomplish His purpose of saving a people for Himself. Therefore, the Lord Jesus incurred the penalty on behalf of God's chosen people - God's elect. These are the people - the *only* people - the Lord Jesus came to save. We read about this in Galatians 3:13-14:

> **Christ hath redeemed us from the curse of the law being made a curse for us: for it is written, Cursed is everyone that hangeth on a tree: That the blessing of Abraham might come on the Gentiles through Jesus Christ; that we might receive the promise of the Spirit through faith.**

The Lord Jesus became a curse for us so that we might escape the curse that had come upon us. Everyone who has sinned is under a curse, as we read in Jeremiah 11:3:

> **And say thou unto them, Thus saith the LORD God of Israel; Cursed *be* the man that obeyeth not the words of his covenant.**

The covenant is God's law. Mankind was accursed by God because we didn't obey the words of the law, just as the ground itself was accursed by God after Adam disobeyed Him. We already read about this in Genesis 3:17:

> **And unto Adam he said, Because thou hast hearkened unto the voice of thy wife, and hast eaten of the tree, of which I commanded thee, saying, Thou shalt not eat of it: cursed *is* the ground for thy sake; in sorrow shalt thou eat *of* it all the days of thy life;**

By cursing the ground, however, God had in a sense cursed mankind because man came from the ground. In Genesis 3:23, we read:

> Therefore the LORD God sent him forth from the garden of Eden, to till the ground from whence he was taken.

At the time Adam and Eve were driven from the garden, all mankind was in a sense "in Adam" because we are all descended from him. The elect are freed from the curse as a result of the Lord's atoning sacrifice. We also read about our freedom from the law's penalty in 2 Corinthians 5:21:

> For he hath made him *to be* sin for us, who knew no sin; that we might be made the righteousness of God in him.

The Lord Jesus had no sins of His own, so the punishment inflicted on Him wasn't needed to pay for any sins He had committed. That punishment could be applied as payment for our sins.

Other Changes to God's Laws

When the Lord Jesus met the requirements of the law (that is, when He fulfilled the law), He brought about changes to the law in addition to those we have already seen. Just as the fourth commandment is no longer binding, and there is no longer a requirement for animal sacrifice, similarly there isn't a requirement for circumcision any more.

Circumcision was given to ancient Israel as a sign. It was a picture of the need for God's people to have sins cut off in order for them to be saved. All males were to have the foreskin of their reproductive organ cut off as a sign of the seed who would come, the seed whose blood would be shed. That seed of course is the Lord Jesus. Circumcision may still make sense for health reasons, but it is no longer a law that God requires His people to follow.

In the early church, there was much contention over the practice of circumcision. The apostle Paul argued forcefully against it in the book of Galatians. In Galatians 5:2-4, we read:

> Behold, I Paul say unto you, that if ye be circumcised, Christ shall profit you nothing. For I testify again to

every man that is circumcised, that he is a debtor to do the whole law. Christ is become of no effect unto you, whosoever of you are justified by the law; ye are fallen from grace.

Of course, Paul isn't saying here that anyone who has already been circumcised is now in trouble with God. He's saying (under the inspiration of God!) that if you now go ahead and let yourself be circumcised because you are trying to follow the law of God, you are putting your trust in your own ability to keep the whole law. But we know that we can't possibly keep the whole law. We can only be saved by the grace of God.

As we read the New Testament, we find that two new ceremonial laws were established. These are water baptism and the Lord's table. We need to keep in mind, however, that we can't be saved by keeping these ceremonial laws anymore than anyone could have been saved by keeping the Old Testament ceremonial law of circumcision.

We read in Matthew 3:13-15 that the Lord Jesus came to John the Baptist to be baptized with water (the word "suffered" means "permitted"):

> **Then cometh Jesus from Galilee to Jordan unto John, to be baptized of him. But John forbad him, saying, I have need to be baptized of thee, and comest thou to me? And Jesus answering said unto him, Suffer** *it to be so* **now: for thus it becometh us to fulfil all righteousness. Then he suffered him.**

And in John 1:33, we read a statement made by John the Baptist about the Lord Jesus, recorded by the apostle John:

> **And I knew him not: but he that sent me to baptize with water, the same said unto me, Upon whom thou shalt see the Spirit descending, and remaining on him, the same is he which baptizeth with the Holy Ghost.**

Baptism with water pictured the washing away of sin that occurs when God baptizes someone with the Holy Spirit. Water baptism was practiced by the early church, and could continue so long as the Holy Spirit was present there. But is the Holy Spirit still present in today's Christian local congregations? We will come back to this question later on.

We read about the establishment of the Lord's table in the Gospel accounts of the Lord's last meal with His apostles. In Luke 22:15-19, we read:

> **And he said unto them, With desire I have desired to eat this passover with with you before I suffer: For I say unto you, I will not any more eat therof, until it be fulfilled in the kingdom of God. And he took the cup, and gave thanks, and said, Take this, and divide *it* among yourselves: For I say unto you, I will not drink of the fruit of the vine, until the kingdom of God shall come. And he took bread, and gave thanks, and brake *it*, and gave unto them, saying, This is my body which is given for you: this do in remembrance of me.**

In Paul's epistles, we find references to this New Testament ceremonial law confirming that it was being kept by the early church when they met together. This ceremonial law was called the Lord's table. We find a description of it in 1 Corinthians 10:16:

> **The cup of blessing which we bless, is it not the communion of the blood of Christ? The bread which we break, is it not the communion of the body of Christ?**

In 1 Corinthians 10:21, it is referred to as the Lord's table:

> **Ye cannot drink the cup of the Lord, and the cup of devils: ye cannot be partakers of the Lord's table, and of the table of devils.**

Like water baptism, the ceremony of the Lord's table was to be practiced by the local congregations. However, it would not apply if ever the time came when God commanded true believers to flee their congregation. We will examine this issue in a later chapter.

No Divorce

The Lord Jesus also made a change to the law when He taught about marriage. In the Old Testament, divorce was permitted, as we read in Deuteronomy 24:1-2:

> **When a man hath taken a wife, and married her, and it come to pass that she find no favour in his eyes, because he hath found some uncleanness in her: then let him write her a bill of divorcement, and give *it* in her hand, and send her out of his house. And when she is departed out of his house, she may go and be another man's *wife*.**

In response to a question from the Pharisees about this matter, the Lord stated that divorce had been permitted "because of the hardness of your hearts," but that it was originally not God's intention to allow divorce. In Luke 16:18, we read a statement that the Lord Jesus made to the Pharisees:

> **Whosoever putteth away his wife, and marrieth another, committeth adultery: and whosoever marrieth her that is put away from *her* husband commiteth adultery.**

This clear teaching stands in stark contrast to the Old Testament scripture that allowed divorce.

The Annual Feasts

Just as God established a weekly Sabbath and made it one of the Ten Commandments, He also established special days - annual feasts - which He commanded ancient Israel to keep. These special days were to be observed as part of God's laws.

The first of these feasts is the Passover. It was to be observed by killing, preparing and eating a lamb, as we read in Deuteronomy 16:2-4:

> **Thou shalt therefore sacrifice the passover unto the LORD thy God, of the flock and the herd, in the place which the LORD shall choose to place his name there.**

> Thou shalt eat no leavened bread with it; seven days shalt thou eat unleavened bread therewith, *even* the bread of affliction: for thou camest forth out of the land of Egypt in haste: that thou mayest remember the day when thou camest forth out of the land of Egypt all the days of thy life. And there shall be no leavened bread seen with thee in all thy coast seven days; neither shall there *any thing* of the flesh, which thou sacrificedst the first day at even, remain all night until the morning.

The Passover was held in the first month (called Nisan) on the fourteenth day according to the calendar given to ancient Israel. It was immediately followed by a seven day feast referred to as the Feast of Unleavened Bread. Twice during the seven days of unleavened bread, all the people were to come together, as we read in Exodus 12:15-16:

> Seven days shall ye eat unleavened bread; even the first day ye shall put away leaven out of your houses: for whosoever eateth leavened bread from the first day until the seventh day, that soul shall be cut off from Israel. And in the first day *there shall be* an holy convocation, and in the seventh day there shall be an holy convocation to you; no manner of work shall be done in them, save *that* which every man must eat, that only may be done of you.

The next feast to occur as the year progressed is known as Pentecost. That's what it's called in the New Testament. Ancient Israel knew it as the Feast of Firstfruits, or the Feast of Weeks, or the Feast of Harvest. This feast was to be kept in the land that ancient Israel would conquer and possess. In Leviticus 23:10-11, we read of a solemn ceremony of the Levitical priesthood:

> Speak unto the children of Israel, and say unto them, When ye be come into the land which I give unto you, and shall reap the harvest thereof, then ye shall bring a sheaf of the firstfruits of your harvest unto the priest: And he shall wave the sheaf before the LORD, to be accepted for you: on the morrow after the sabbath the

priest shall wave it.

The day on which Pentecost was to be observed was determined by counting days following the wave sheaf offering, as we read in Leviticus 23:15-16:

> **And ye shall count unto you from the morrow after the sabbath, from the day that ye brought the sheaf of the wave offering; seven sabbaths shall be complete: Even unto the morrow after the seventh sabbath shall ye number fifty days; and ye shall offer a new meat offering unto the LORD.**

The name "Pentecost" comes from the Greek word for fiftieth. Pentecost is the only one of these annual feast that was determined by counting days. All the others were determined by Israel's calendar.

As we read through the twenty third chapter of Leviticus, the next feast we come to is called the Feast of Trumpets. We read in Leviticus 23:24 that it was to be celebrated by a blowing of trumpets:

> **Speak unto the children of Israel, saying, In the seventh month, in the first *day* of the month, shall ye have a sabbath, a memorial of blowing of trumpets, an holy convocation.**

For obvious reasons, this feast has come to be known as the Feast of Trumpets. However, if you were to check on the original Hebrew word used in this scripture, translated as "blowing of trumpets," you would find that it is a word ("teruah") that *does not* refer to the musical instrument we know as the trumpet. Rather, it refers to the Jubilee - a special year that occurred every fifty years and was signaled by a blast on the ram's horn on the Day of Atonement (the next annual feast in ancient Israel's calendar of feast days). The Feast of Trumpets occurred in the seventh month on the first day of the month - at the new moon. Instead of calling this the Feast of Trumpets, we could call it the Feast of Jubilee.

The next annual feast listed in the twenty third chapter of

Leviticus, which lists all the annual feasts, is the Day of Atonement. This feast also took place in the seventh month. In Leviticus 23:27, we read:

> **Also on the tenth *day* of this seventh month *there shall be* a day of atonement: it shall be an holy convocation unto you; and ye shall afflict your souls, and offer an offering made by fire unto the LORD.**

As stated earlier, once every fifty years on this day the ram's horn was sounded to mark the year of Jubilee. Later on, we will take a closer look at the Jubilee.

Shortly after the Day of Atonement, the Feast of Tabernacles took place. The description of this feast begins in Leviticus 23:34:

> **Speak unto the children of Israel, saying, The fifteenth day of this seventh month *shall be* the feast of tabernacles *for* seven days unto the LORD.**

This feast began with a day on which servile work was forbidden; it was followed by seven days of sacrifices, and ended with another day on which servile work was forbidden. We read about this in Leviticus 23:35-36:

> **On the first day *shall be* an holy convocation: ye shall do no servile work *therein*. Seven days ye shall offer an offering made by fire unto the LORD: on the eighth day shall be an holy convocation unto you; and ye shall offer an offering made by fire unto the LORD: it *is* a solemn assembly; *and* ye shall do no servile work *therein*.**

The Feast of Tabernacles is the last of the feasts listed in this part of the Bible. In Leviticus 23:37-38, we read:

> **These *are* the feasts of the LORD, which ye shall proclaim *to be* holy convocations, to offer an offering made by fire unto the LORD, a burnt offering, and a meat offering, a sacrifice, and drink offerings, every thing upon his day: Beside the sabbaths of the LORD, and beside your gifts, and beside all your vows, and**

beside all your freewill offerings, which ye give unto the LORD.

Just as the requirements for animal sacrifices and the keeping of the seventh day sabbath are no longer in force, we aren't required to keep these annual feasts either. Why then should we be concerned about these annual feasts? We will come back to this subject in the next chapter.

The Sabbath Year

In addition to the annual feasts, every seventh year was special to ancient Israel. In Leviticus 25:3-4, we read:

> Six years thou shalt sow thy field, and six years thou shalt prune thy vineyard, and gather in the fruit thereof; But in the seventh year shall be a sabbath of rest unto the land, a sabbath for the LORD: thou shalt neither sow thy field, nor prune thy vineyard.

Israel began to keep Sabbath years when they first entered the land of Canaan, their first year in the land being the first Sabbath year. We may compare the Sabbath year to the Sabbath day. When Israel was in the wilderness and the Lord was feeding them with manna, the people were not to go looking for manna on the Sabbath day. The Lord provided them with an extra portion on Friday, as we read in Exodus 16:29:

> See, for that the LORD hath given you the sabbath, therefore he giveth you on the sixth day the bread of two days; abide ye every man in his place, let no man go out of his place on the seventh day.

Similarly, the Lord promised to provide for His people even though they did not plant their crops or prune their vineyards every seventh year. In Leviticus 25:20-21, we read:

> And if ye shall say, What shall we eat the seventh year? behold, we shall not sow, nor gather in our increase: Then I will command my blessing upon you in the sixth

year, and it shall bring forth fruit for three years.

We also see the Sabbath year being compared with food itself, as we read in Leviticus 25:6:

> **And the sabbath of the land shall be meat for you; for thee, and for thy servant, and for thy maid, and for thy hired servant, and for thy stranger that sojourneth with thee,**

This may bring to mind the way the Lord Jesus compared His own body to food, as we read in John 6:54-56:

> **Whoso eateth my flesh, and drinketh my blood, hath eternal life; and I will raise him up at the last day. For my flesh is meat indeed, and my blood is drink indeed. He that eateth my flesh, and drinketh my blood, dwelleth in me, and I in him.**

When we read the Bible, we occasionally find God using ugly language, as He does here, to convey spiritual truth. The seventh year Sabbath was a picture or a shadow of the Lord Jesus. After He had come, we had the substance of what the Sabbath year pictured; and so the keeping of the Sabbath year is no longer required.

The Jubilee Year

The people of ancient Israel were commanded to count Sabbath years in order to determine when to observe a special year known as the Jubilee, as in Leviticus 25:8-9:

> **And thou shalt number seven sabbaths of years unto thee, seven times seven years; and the space of the seven sabbaths of years shall be unto thee forty and nine years. Then shalt thou cause the trumpet of the jubile to sound on the tenth *day* of the seventh month, in the day of atonement shall ye make the trumpet sound throughout all your land.**

Like the Sabbath year, the Jubilee was a year in which the people were commanded to not sow their crops; but there was much greater significance to the Jubilee. The Jubilee year was very special. It was a year of liberty for people and land. If a man had become poor and had sold himself to be a servant, he was to be set at liberty in the jubilee, as we read in Leviticus 25:39-42:

> **And if thy brother *that dwelleth* by thee be waxen poor, and be sold unto thee; thou shalt not compel him to serve as a bondservant: *But* as an hired servant, *and* as a sojourner, he shall be with thee, *and* shall serve thee unto the year of jubile: And *then* shall he depart from thee, *both* he and his children with him, and shall return unto his own family, and unto the possession of his fathers shall he return. For they *are* my servants, which I brought forth out of the land of Egypt: they shall not be sold as bondmen.**

Leviticus 25:10 also tells us what happened in the Jubilee year:

> **And ye shall hallow the fiftieth year, and proclaim liberty throughout *all* the land unto all the inhabitants thereof: it shall be a jubile unto you; and ye shall return every man unto his possession, and ye shall return every man unto his family.**

When we read the New Testament, we find that the Lord Jesus identifies His ministry with the year of Jubilee. We see this when we read Luke 4:18-19:

> **The Spirit of the Lord *is* upon me, because he hath anointed me to preach the gospel to the poor; he hath sent me to heal the brokenhearted, to preach deliverance to the captives, and recovering of sight to the blind, to set at liberty them that are bruised, To preach the acceptable year of the Lord.**

In saying that He was sent to "set at liberty them that are bruised," we should see that the Lord is making a comparison with the requirement in ancient Israel that people be freed from servitude in the Jubilee year. Similarly, in the words "to preach the acceptable year of the Lord" we see a reference to the Jubilee. Later on, we will see that there is great significance in those words.

Many Other Laws

There are many other laws throughout the Bible. Some of these are stated in a way that makes it obvious that they are laws. These verses may, for example, use the words "thou shalt" or "thou shalt not;" but in many other instances, even though verses are not stated obviously as laws, those verses are laws just the same.

As an example of one such law, read 1 Timothy 2:1-3:

> **I exhort therefore, that, first of all, supplications, prayers, intercessions, *and* giving of thanks, be made for all men; For kings, and *for* all that are in authority; that we may lead a quiet and peaceable life in all godliness and honesty. For this *is* good and acceptable in the sight of God our Saviour;**

Since these are God's words, we can't take them as a suggestion or a recommendation. God is telling us that we are to pray for our rulers. Whether our ruler's title is president, king, or something else; whether or not we like our ruler and the way he rules; whether he is just and competent or a thief and tyrant, we are to pray for him.

Some Biblical laws may have implications far beyond any situations or conditions stated in them. In Deuteronomy 25:1-3, we read something that seems clear enough:

> **If there be a controversy between men, and they come unto judgment, that *the judges* may judge them; then they shall justify the righteous, and condemn the wicked. And it shall be, if the wicked man *be* worthy to be beaten, that the judge shall cause him to lie down, and to be beaten before his face, according to his fault, by a certain number. Forty stripes he may give him, *and* not exceed: lest, *if* he should exceed, and beat him above these with many stripes, then thy brother should seem**

vile unto thee.

In this law God gave to ancient Israel, God is obviously setting a limit on the degree to which a man may be punished if he is found guilty of some crime or other. However, when we remember that the Bible is also very concerned with the punishment man incurs when he comes under God's wrath, and when we remember that the Lord Jesus spoke in parables, we realize that God is telling us in these verses that He will not punish anyone without limit. There cannot be a place called Hell in which the unsaved are punished for all eternity, because that would contradict God's law setting a limit on a man's punishment.

Finally, here is an example of a scripture that is not obviously stated as a law, yet it is giving us a very important law. Also, its scope is far reaching going well beyond the circumstances with which it is apparently concerned. Earlier, we took a look at Matthew 28:1, which we will reconsider here:

> **In the end of the sabbath, as it began to dawn toward the first *day* of the week, came Mary Magdalene and the other Mary to see the sepulchre.**

We learned that this verse was not well translated in the King James Bible. A better translation would have been:

> In the end of the Sabbaths, as it began to dawn toward the first of the Sabbaths, came Mary Magdalene and the other Mary to see the sepulchre.

If you search the Bible, especially the New Testament, you won't find a law clearly stating that the seventh day (Saturday) Sabbath is no longer in force, and that today's true believers are to observe the first day of the week as their Sabbath. In Matthew 28:1, however, God is telling us just that. This scripture tells us that the era of the seventh day Sabbaths is over (the end of the Sabbaths) and that a new era of Sunday Sabbaths (the first of the Sabbaths) has begun.

As you read the Bible carefully and with prayer, comparing scripture with scripture, you may find other verses in which God's word opens up to you so that you will understand something you never understood before - even though you may have read those same verses many times.

Chapter 4

God's Problem

It may seem strange to hear that God has a problem. After all, He's God and He can do anything He wants to do. How could He possibly have a problem? Well, He does have a problem, and we're it. You and I, and everyone who has ever lived and who will ever live - we are His problem.

He knew what He was getting into before He even created man. He knew we would reject Him and choose to go our own way. He knew we would grieve Him time and time again, ultimately bringing the terrible penalty of His law on ourselves. The grief He has experienced over us is, however, a result of His love. If He didn't care about any of us, then we would not matter to Him.

Before creating the universe, God had already created angels and perhaps many other types of beings as well. He now wanted to create a new kind of being. This new creation, man, would be made in God's own image and likeness. Man would be special. God created man in order to save some of us for eternal life, and He has given man the Bible to reveal these things to us.

God wants us to know about ourselves and our desperate situation. Scriptures that provide a glimpse of ourselves in relation to God help us to see ourselves as God sees us. King David understood something about this in his wonder at God's creation, as we read in Psalm 8:3-4:

> **When I consider thy heavens, the work of thy fingers, the moon and the stars, which thou hast ordained; What is man, that thou art mindful of him? and the son of**

man, that thou visitest him?

If we begin to see ourselves as God sees us, we may gain some understanding of our need for His mercy.

Man's Tendency

The Bible shows us that man has a tendency to reject God. Man wants to be in charge! We see this in the account of Adam and Eve in the garden of Eden: they rejected God's authority over them when they decided to eat of the tree of the knowledge of good and evil. We also see this illustrated when ancient Israel insisted on having a king over them, in the manner of the other nations at that time. In 1 Samuel 8:7, the Lord is talking to the prophet Samuel:

> **And the LORD said unto Samuel, Hearken unto the voice of the people in all that they say unto thee: for they have not rejected thee, but they have rejected me, that I should not reign over them.**

We also see this tendency in the New Testament in the rejection of the Lord Jesus - the same Lord of the Old Testament who was first rejected by Adam and Eve, then later rejected by ancient Israel. In Luke 19, the Lord Jesus tells a parable about a nobleman who went to a far country to receive a kingdom. In Luke 19:14, we read:

> **But his citizens hated him, and sent a message after him, saying, We will not have this *man* to reign over us.**

The parable ends in Luke 19:27, where we read what the king eventually commands:

> **But those mine enemies, which would not that I should reign over them, bring hither, and slay *them* before me.**

It's seems so easy for man to focus entirely on himself and his own gratification, without ever considering God or seeking His will. In fact, it seems to be in our nature to do this. The results of this way of life are what the Bible calls the works of the flesh. In Galatians 5:19-21, we read:

Chapter 4 God's Problem

> Now the works of the flesh are manifest, which are *these*; Adultery, fornication, uncleanness, lasciviousness, Idolatry, witchcraft, hatred, variance, emulations, wrath, strife, seditions, heresies, Envyings, murders, drunkenness, revellings, and such like: of the which I tell you before, as I have also told *you* in time past, that they which do such things shall not inherit the kingdom of God.

These conditions plague our world because we have forgotten God. Yet God tells us that He is to be central in our lives. In Deuteronomy 6:4-5, we read:

> Hear, O Israel: The LORD our God *is* one LORD: And thou shalt love the LORD thy God with all thine heart, and with all thy soul, and with all thy might.

This command to love God is repeated in the New Testament. In Matthew 22:36-37, one of the Pharisees is questioning Jesus:

> Master, which *is* the great commandment in the law? Jesus said unto him, Thou shalt love the Lord thy God with all thy heart, and with all thy soul, and with all thy mind.

On his own, man wants to exclude God from his life - despite this command and the suffering that clearly results from a life that is not led by God. Man's situation is, however, even worse than this: the Bible reveals that man has a deadly enemy.

Man's Enemy

God wants us to know that Satan, also called the devil, is our deadly enemy. One of the Bible's references to him is 1 Peter 5:8:

> Be sober, be vigilant; because your adversary the devil, as a roaring lion, walketh about, seeking whom he may devour:

How did this terrible being come to exist? Incredibly, we learn from the Bible that God created him! Satan is an angel who rebelled against God and led other angels to rebel as well. God created all the angels to serve Him, and He created Satan too. But Satan didn't want to serve God - he wanted to be in control in place of God. The Lord Jesus spoke of him in John 8:44:

> **Ye are of *your* father the devil, and the lusts of your father ye will do. He was a murderer from the beginning, and abode not in the truth, because there is no truth in him. When he speaketh a lie, he speaketh of his own: for he is a liar, and the father of it.**

We learn something else about the devil in Ephesians 2. It is true that the Bible is for all mankind, but many verses are addressed only to true believers. Anyone may read them and hopefully benefit by them, but what they speak of is true only of those God has already saved. Ephesians 2:1-3 are such verses, and also provide information about Satan. In Ephesians 2:1-3, we read:

> **And you *hath he quickened*, who were dead in trespasses and sins; Wherein in time past ye walked according to the course of this world, according to the prince of the power of the air, the spirit that now worketh in the children of disobedience: Among whom also we all had our conversation in times past in the lusts of our flesh, fulfilling the desires of the flesh and of the mind; and were by nature the children of wrath, even as others.**

"The prince of the power of the air" is Satan. Perhaps this scripture is telling us that he can broadcast his God-opposing, hateful thoughts - almost like a radio station sends its signal out over the air - to influence and possibly even control people. We read in these verses that Satan's spirit is at work in us unless God has given us new life ("quickened" us).

In addition to calling him a roaring lion, the Bible also calls Satan a dragon in Revelation 12. In Revelation 12:9, we read:

> **And the great dragon was cast out, that old serpent called the Devil, and Satan, which deceiveth the whole world: he was cast out into the earth, and his angels**

were cast out with him.

Putting all these verses together, we see that mankind is being deceived by Satan, and that he is a powerful and influential enemy who is actually stalking us. Satan was involved in the downfall of Adam and Eve, and he has been the enemy of mankind ever since. God has warned us about him, and God will hold man accountable - despite Satan's work against us - just as He held Adam and Eve accountable.

Man's Preoccupation

God does not want us to be totally preoccupied with ourselves and our world. Unfortunately, it seems that most of us are. Satan undoubtedly works to create this condition because it means our focus won't be on God. For so many people around the world throughout history, and even today, a life of poverty requires that they work many long hours just to survive. In the modern industrialized nations, where a large middle class exists, many people have some discretionary funds and leisure time. For those of us who are so blessed, it seems too easy to forget about God. The world has a great many alluring forms of entertainment. You may be a sports fan and spend all your spare time watching sports on television. Maybe you are immersed in playing video games. Perhaps you can afford some of the finer things of life - it's easy to become preoccupied in chasing the world's material goods. You may possibly have become trapped in an addictive form of behavior. All of these things keep our attention away from God. God warns us about the danger of being preoccupied with the world. In 1 John 2:15-16, we read:

> **Love not the world, neither the things *that are* in the world. If any man love the world, the love of the Father is not in him. For all that *is* in the world, the lust of the flesh, and the lust of the eyes, and the pride of life, is not of the Father, but is of the world.**

Interestingly, we see those same three lures of the world - the lust of the flesh, the lust of the eyes, and the pride of life - at work in

Eve in the garden of Eden. In Genesis 3:6, when she was about to break God's command not to eat of the one tree forbidden to her and Adam, we read of Eve:

> **And when the woman saw that the tree *was* good for food, and that it *was* pleasant to the eyes, and a tree to be desired to make *one* wise, she took of the fruit thereof, and did eat, and gave also unto her husband with her; and he did eat.**

Eve saw that the tree was good for food (the lust of the flesh), that it was pleasant to the eyes (the lust of the eyes), and that it was to be desired to make one wise (the pride of life). Human nature hasn't changed since then. That's why the Bible is a book for us today just as much as it has been for mankind throughout the ages.

It may seem perfectly natural for us to forget about God, to be entirely wrapped up in our families, careers, or our own pleasures; but the Bible tells us that we are to love God and to obey His commands. As we try to do His will, we must always remember that nothing we do can earn salvation for us. That is a trap that many people have fallen into, as we are warned in Proverbs 14:12:

> **There is a way which seemeth right unto a man, but the end therof *are* the ways of death.**

Man's Fallen Nature

It seems natural for us to live without acknowledging God, because that is our nature. The Bible makes some shocking statements about our nature, and our natural inclination is to refuse to believe what it reveals. For example, we learn from Jeremiah 17:9:

> **The heart *is* deceitful above all *things*, and desperately wicked: who can know it?**

In the Psalms, there is more about this. We read in Psalm 14:2-3:

> **The LORD looked down from heaven upon the children**

of men, to see if there were any that did understand, *and seek God. They are all gone aside, they are all together become filthy: there is none that doeth good, no, not one.*

There is a similar scripture in the New Testament. In Romans 3:10-11, we read:

As it is written, There is none righteous, no, not one: There is none that understandeth, there is none that seeketh after God.

We may think of the sin of Adam and Eve as an infection that has spread through all succeeding generations to the entire human race, almost as if it's embedded in our genetic makeup. Interestingly, the Bible has a number of scriptures concerned with the terrible disease known as leprosy. In these scriptures, God may be showing us how He sees sin: leprosy is a picture of sin.

You might reason that God wouldn't consider a little baby to be responsible for sin - that a person could only be held accountable after reaching a certain age; but the Bible tells us that our fallen nature manifests itself from the start of our lives. Read what God says about the wicked in Psalm 58:3-4:

The wicked are estranged from the womb: they go astray as soon as they be born, speaking lies. Their poison *is* like the poison of a serpent: *they are* like the deaf adder *that* stoppeth her ear;

According to this scripture, God considers that cute, innocent, baby to be like a poisonous snake - a little adder. We love ourselves more than truth, and we don't want to hear anything that contradicts our proud self-image resulting from our fallen nature.

Man and Music

Just as God gave man the gift of language, He has also given us the gift of music. Sadly, both of these have been corrupted by sin.

As we read the Bible, we learn that God's purpose in giving man the ability to create and enjoy music is that the music should bring glory to God. Surely, today's music has deviated from that purpose more than ever before. Much of today's music (and art) seems to be concerned with criminal activity, with sensuality, with decay and death, and even with bodily functions. Some music mocks the Bible and mocks God sometimes in a very obvious and purposeful way. Even music with lyrics about life's typical concerns, whether serious or trivial - music that really cannot be considered offensive in any way - even this kind of music is a misuse of our God-given abilities.

Some local congregations and other Christian organizations have argued that worldly music has its place in worship services because that is the kind of music that many people, especially young people, enjoy today; so they use such music, with lyrics they consider appropriate for the Gospel message, hoping to increase and retain their membership. The trouble is, just as vile lyrics flow from the human heart, so does the kind of music that most of this world seems to enjoy. Ephesians 5:19-20 give us some guidance in this matter, telling us how we ought to conduct ourselves:

> **Speaking to yourselves in psalms and hymns and spiritual songs, singing and making melody in your heart to the Lord; Giving thanks always for all things unto God and the Father in the name of our Lord Jesus Christ;**

If someone is giving thanks to God at all times, "singing and making melody" in his heart to the Lord, then there will be no time in that person's life for any music other than psalms, hymns and spiritual songs. (If we were to really follow it, this Biblical injunction would also rule out most other forms of worldly entertainment, such as sports and video games.)

Even a composer with no intention of glorifying God can write beautiful music suitable for worship, although the lyrics for such music may have to be completely rewritten. Beethoven's Ode to Joy from his Ninth Symphony is an example. Although the lyrics used for this music do mention God, they are far from being Biblical; yet the music is joyful and inspirational, and is currently used with new lyrics that offer praise and thanksgiving to God.

The Book of Psalms includes a number of verses dealing with the use of music to praise God. Psalm 149:2-3 is one such example:

> **Let Israel rejoice in him that made him: let the children of Zion be joyful in their King. Let them praise his name in the dance: let them sing praises unto him with the timbrel and harp.**

Here we see mention of both dance and music to praise God. Just as the purpose of music is to glorify God, so it is for dance and all other art forms. Sadly, only a tiny fraction of all music, dance and art fulfills God's intended purpose.

When God's elect are complete - immortal body united with immortal soul - their music will be what God always intended. It will be music that glorifies God: full of joy, thanksgiving, praise and worship.

Man and Alcoholic Beverages

The Bible instructs God's elect not to drink wine or strong drink. For anyone who is acquainted with the Bible, this claim may seem incorrect. After all, the changing of water into wine was the first public miracle performed by the Lord Jesus; and the apostle Paul even wrote to his helper Timothy, telling him to drink a little wine. Why would the Lord make all that additional wine available to the wedding guests if God didn't want them to drink it, and why should Paul encourage anyone to drink wine if it's wrong to do so? It turns out that it is only God's elect who are not to drink wine or strong drink.

The Bible's warning against becoming drunk is obvious. In 1 Corinthians 6:9-10, we read:

Know ye not that the unrighteous shall not inherit the kingdom of God? Be not deceived: neither fornicators, nor idolaters, nor adulterers, nor effeminate, nor abusers of themselves with mankind, Nor thieves, nor covetous, nor drunkards, nor revilers, nor extortioners, shall inherit the kingdom of God.

There, in the list of persons practicing all types of sinful behavior, we plainly read that no drunkard shall inherit the kingdom of God Of course, there's a big difference between drinking and being a drunkard; but in order to understand what the Bible is teaching on this matter, we need to examine additional scriptures.

First, let's read Proverbs 31:4:

***It is* not for kings, O Lemuel, *it is* not for kings to drink wine; nor for princes strong drink:**

Another scripture we should read is Ezekiel 44:21:

Neither shall any priest drink wine, when they enter into the inner court.

A priest who had entered into the inner court had moved closer to the most holy place in the temple, and he was not to drink wine The king was not to drink wine at any time. From the two preceding scriptures, we see prohibitions against drinking for both priests and kings. Let us now consider Revelation 1:6, where we read the following:

And hath made us kings and priests unto God and his Father; to him *be* glory and dominion for ever and ever. Amen.

God sees His elect as kings and priests. They are always in the "inner court" so to speak, and so they are not to drink wine or alcoholic beverages. There are exceptions for health, and this was Paul's advice to Timothy in 1 Timothy 5:23:

Drink no longer water, but use a little wine for thy stomach's sake and thine often infirmities.

Timothy knew he shouldn't be drinking wine, and so he didn't. Paul is telling Timothy in this verse that the medicinal use of wine is all right. Today, God's elect should not be "social drinkers" or any other kind, even if they can drink without being tempted to drink to excess. However, wine in small amounts for medicinal use is fine.

Man and Government

In the more than 1900 years since the Bible has been completed, it has surely been quoted many times to support the overthrow of tyrannical governments. Does the Bible really tell people that they should or may revolt against their government under certain conditions? No, it doesn't. On the contrary, God's word is very friendly to every government, no matter how incompetent, how cruel, or how unjust it is to its people. Take, for example, the Biblical teaching concerning the payment of taxes. In an attempt to trap the Lord Jesus into saying something that could be used against Him, the religious leaders of His day asked Him about paying taxes. In Mark 12:14, we read their question:

> **And when they were come, they say unto him, Master, we know that thou art true, and carest for no man: for thou regardest not the person of men, but teachest the way of God in truth: Is it lawful to give tribute to Caesar, or not?**

Mark 12:15-17 continues with their discussion and the Lord's response to that question:

> **Shall we give, or shall we not give? But he, knowing their hypocrisy, said unto them, Why tempt ye me? bring me a penny, that I may see** *it*. **And they brought** *it*. **And he saith unto them, Whose** *is* **this image and superscription? And they said unto him, Caesar's. And Jesus answering said unto them, Render to Caesar the things that are Caesar's, and to God the things that are God's. And they marvelled at him.**

Those who are trying to live by the word of God are instructed in these verses to pay their lawful taxes. The Bible is supportive of established rule in other verses as well. In Daniel 2:21, we read part of Daniel's prayer blessing the God of heaven and telling of things He does:

> **And he changeth the times and the seasons: he removeth kings, and setteth up kings: he giveth wisdom unto the wise, and knowledge to them that know understanding:**

If a king or any other kind of ruler or person in authority is to be replaced, God will do it. We may wish that someone else were ruling over us, but it is God who is in charge of these matters.

Man is Special

Even though many scriptures are extremely negative about mankind, we learn that God treats human beings differently than all other creatures because we are special. The very fact that God gave us the Bible proves this. Also, we know that we are somehow patterned after God Himself, as we read in Genesis 1:26:

> **And God said, Let us make man in our image, after our likeness: and let them have dominion over the fish of the sea, and over the fowl of the air, and over the cattle, and over all the earth, and over every creeping thing that creepeth upon the earth.**

God knows each one of us individually. The fact that He has recorded for us the names of so many individuals illustrates this. We also know that God gave man a conscience. We read about the conscience in Romans 2:13-15:

> **(For not the hearers of the law *are* just before God, but the doers of the law shall be justified. For when the Gentiles, which have not the law, do by nature the things contained in the law, these, having not the law, are a law unto themselves: Which shew the work of the law written in their hearts, their conscience also bearing witness, and *their* thoughts the mean while accusing or else excusing one another;)**

We see the conscience at work in the hearts of the scribes and Pharisees in an incident recorded in the Gospel According to John. In John 8:5-9, a group of them is about to stone a woman for adultery and they are speaking to the Lord Jesus:

> **Now Moses in the law commanded us, that such should be stoned: but what sayest thou? This they said, tempting him, that they might have to accuse him. But Jesus stooped down, and with *his* finger wrote on the ground, *as though he heard them not*. So when they continued asking him, he lifted up himself, and said unto them, He that is without sin among you, let him first cast a stone at her. And again he stooped down, and wrote on the ground. And they which heard *it*, being convicted by *their own* conscience, went out one by one, beginning at the eldest, *even* unto the last: and Jesus was left alone, and the woman standing in the midst.**

From the Gospel accounts, it certainly seems that most of the scribes and Pharisees were never saved. Yet here we see a group of them convicted by their conscience. Perhaps it is because we are made in the image and likeness of God Himself that He holds man accountable for sin; and because we are accountable, there is a penalty for sin.

God Grieves

God tells us that He will take no pleasure in exacting this penalty. In Ezekiel 18:32, we read:

> **For I have no pleasure in the death of him that dieth, saith the Lord GOD: wherefore turn *yourselves*, and live ye.**

It grieved God to destroy mankind in the great flood that covered the whole earth thousands of years ago. Except for Noah and his family, all perished. We read in Genesis 6:5-6:

> And God saw that the wickedness of man *was* great in the earth, and *that* every imagination of the thoughts of his heart *was* only evil continually. And it repented the LORD that he had made man on the earth, and it grieved him at his heart.

In the New Testament, we also read that God doesn't want to see anyone perish, as stated in 2 Peter 3:9:

> The Lord is not slack concerning his promise, as some men count slackness; but is longsuffering to us-ward, not willing that any should perish, but that all should come to repentance.

Also in the New Testament, we learn that Jesus wept over Jerusalem. In Luke 19:41-42, we read:

> And when he was come near, he beheld the city, and wept over it, Saying, If thou hadst known, even thou, at least in this thy day, the things *which belong* unto thy peace! but now they are hid from thine eyes.

Many people who are familiar with this passage will claim that the reason Jesus wept over Jerusalem is that He knew of the city's coming destruction by the Romans. This certainly seems to be the obvious reason; but it is not necessarily the only reason Jesus wept. The obvious meaning of a scripture doesn't have to be the only meaning. Just as the word "death" in Ezekiel 18:32 refers to the penalty for sin - what the Bible elsewhere calls the second death - the city of Jerusalem in scripture represents more than just the earthly city. The Lord Jesus wept over those who must suffer the second death when He wept over Jerusalem, just as God grieved over those who were going to suffer the second death when He was about to destroy mankind in the flood. The second death will occur on the last day of the earth's existence, when the unsaved who are still alive are annihilated along with the remains of all those unsaved persons who have died since the creation. God knows what is coming because He is all powerful to execute His plan. He has been grieving about this matter for thousands of years.

God created man to enjoy a wonderful, eternal inheritance;

but when someone who is not one of God's elect dies, that inheritance is lost forever. That person will never live again: he or she has been denied eternal life with the Lord and all the beauty and joy that go with it. On the other hand, when someone who is one of God's children dies, he or she goes to be with Him. That person's death is a transition to a new existence that will include inheriting the new heaven and the new earth that God will soon create.

The Fear of the Lord

God has given us the Bible to let us know that we are in trouble with Him. He wants us to realize that we are in a desperate situation, and to cry out to Him for mercy. In Psalm 111:10, we read:

> **The fear of the LORD *is* the beginning of wisdom: a good understanding have all they that do *his commandments*: his praise endureth for ever.**

There is an incident in the Book of 2 Chronicles about the discovery of the book of the law in the temple during the reign of Josiah. In 2 Chronicles 34:19-21, we see how the king reacted for fear of God's wrath when the book was read before him:

> **And it came to pass, when the king had heard the words of the law, that he rent his clothes. And the king commanded Hilkiah, and Ahikam the son of Shaphan, and Abdon the son of Micah, and Shaphan the scribe, and Asaiah a servant of the king's, saying, Go, inquire of the LORD for me, and for them that are left in Israel and in Judah, concerning the words of the book that is found: for great *is* the wrath of the LORD that is poured out upon us, because our fathers have not kept the word of the LORD, to do after all that is written in this book.**

We know that king Josiah was a child of God. Perhaps God has caused this incident to be recorded and preserved for us as an example of true repentance.

In the wilderness, the children of Israel also experienced great fear when they witnessed a display of God's power at mount Sinai after they had come out of Egypt. We read about this in Exodus 20:18-20:

> **And all the people saw the thunderings, and the lightnings, and the noise of the trumpet, and the mountain smoking: and when the people saw *it*, they removed, and stood afar off. And they said unto Moses, Speak thou with us, and we will hear: but let not God speak with us, lest we die. And Moses said unto the people, Fear not: for God is come to prove you, and that his fear may be before your faces, that ye sin not.**

Despite this event and many others in which God showed them a glimpse of His power, Moses' hope for his people was not realized. Practically the entire generation that had escaped from Egypt perished in the wilderness. They are a picture for us of people who were never saved.

In the book of Proverbs, there are many scriptures warning us to fear God. Many of them may just appear to be sayings that illustrate common sense or moral truth rather than any spiritual meaning. In the case of Proverbs 1:27-30, the warning to fear the Lord is clearly stated:

> **When your fear cometh as desolation, and your destruction cometh as a whirlwind; when distress and anguish cometh upon you. Then shall they call upon me, but I will not answer; they shall seek me early, but they shall not find me: For that they hated knowledge, and did not choose the fear of the LORD: They would none of my counsel: they despised all my reproof.**

Such warnings appear throughout the Bible. An example from the New Testament is Matthew 10:28, in which the Lord Jesus warned His disciples what to fear:

> And fear not them which kill the body, but are not able to kill the soul: but rather fear him which is able to destroy both soul and body in hell.

This scripture too is a warning to fear God, and the person the Lord is warning us to fear is Himself - it is He who will come to judge the living and the dead.

We might not even realize we're in trouble with God, but God tells us that in our sinful lives we are actually His enemies. We see this in Romans 5:10:

> For if, when we were enemies, we were reconciled to God by the death of his Son, much more, being reconciled, we shall be saved by his life.

We read something similar in Colossians 1:21-22:

> And you, that were sometime alienated and enemies in *your* mind by wicked works, yet now hath he reconciled In the body of his flesh through death, to present you holy and unblameable and unreproveable in his sight:

Unless we are reconciled to God, we're heading for a terrible fate. Hebrews 10:31 tells us:

> *It is* a fearful thing to fall into the hands of the living God.

What then are we to do once we realize that we have incurred the penalty for sin, and that we will eventually face judgment? The Lord Jesus told a parable about two men who prayed to God. In Luke 18:10, we read:

> Two men went up into the temple to pray; the one a Pharisee, and the other a publican.

The Pharisees were knowledgeable (or so they thought) about the scriptures, and probably considered themselves to be right with God. The publicans were tax collectors, and had the reputation of being less than honest in the commission of their duties. The story continues in Luke 18:11-13:

> The Pharisee stood and prayed thus with himself, God, I thank thee, that I am not as other men *are*, extortioners, unjust, adulterers, or even as this publican. I fast twice in the week, I give tithes of all that I possess. And the publican, standing afar off, would not lift up so much as *his* eyes unto heaven, but smote upon his breast, saying, God be merciful to me a sinner.

In the next verse, Luke 18:14, the Lord Jesus tells us how God sees these two men:

> I tell you, this man went down to his house justified *rather* than the other: for every one that exalteth himself shall be abased; and he that humbleth himself shall be exalted.

We should pray as the publican did, crying out to God for mercy and asking Him to save us. God tells us that He is merciful, and that this is still the day of salvation. As long as you are alive, until the Lord returns, there is hope you may be saved.

The Saved

How did God choose those who will be saved? This is something we don't know - it is God's business. Since this decision was made before the foundation of the world, we know that God looked down through time - into the future to the end of time - at all people who would ever live. From all the billions of people all over the planet and throughout history, God made His choice. Those He chose to save, after they die, go to be with the Lord in their spirit. This is where they are and will remain until the Lord returns.

We know that God sees people differently than we do. People see each other through eyes that have been trained to notice social, economic and other distinctions. We judge people in this way, and tend to treat them accordingly. Frequently, we are less than respectful to the poor. An example - and warning - for us has been recorded in the second chapter of James. Some members of the early church were guilty of this sort of behavior: that is, of favoring the wealthy at the expense of the poor. In James 2:2-4, we read:

> **For if there come unto your assembly a man with a gold ring, in goodly apparel, and there come in also a poor man in vile raiment; And ye have respect to him that weareth the gay clothing, and say unto him, Sit thou here in a good place; and say to the poor, Stand thou there, or sit here under my footstool: Are ye not then partial in yourselves, and are become judges of evil thoughts?**

God instructs us not to treat people based on how much money they have. Neither are we to favor anyone on the basis of their speech, race, nationality, educational level, etc. In James 2:8-10, we read:

> **If ye fulfil the royal law according to the scripture, Thou shalt love thy neighbour as thyself, ye do well: But if ye have respect to persons, ye commit sin, and are convinced of the law as transgressors. For whosoever shall keep the whole law, and yet offend in one *point*, he is guilty of all.**

God tells us that He is no respecter of persons. He isn't impressed by how wealthy you are, how popular you are, how smart you are, or anything else that impresses people. Every human is so far beneath God that anything we can accomplish is no big deal to Him. Besides, He gave every one of us the means to do every good thing we can ever do. He has even given us our very life and any special gifts we may possess.

The people who will be saved need not be outstanding in any way, so there is great hope for each one of us. In Deuteronomy 10:17-18, we read:

> **For the LORD your God *is* God of gods, and Lord of lords, a great God, a mighty, and a terrible, which regardeth not persons, nor taketh reward: He doth execute the judgment of the fatherless and widow, and**

loveth the stranger, in giving him food and raiment.

In Acts 10:34-35, we read more about this in a statement made by the apostle Peter to a soldier named Cornelius:

Then Peter opened *his* mouth, and said, Of a truth I perceive that God is no respecter of persons: But in every nation he that feareth him, and worketh righteousness, is accepted with him.

Those who have been saved can truly love God, as we read in 1 John 4:19:

We love him, because he first loved us.

If anyone can love God, it is only because God first loved him! And if God loves anyone, that person is saved. The saved are often referred to as the "elect," as in 2 John 1:1, where we read:

The elder unto the elect lady and her children, whom I love in the truth; and not I only, but also all they that have known the truth;

God tells us that He is concerned about His glory, and He will reveal His glory through His elect. Many of these people have been chosen from the humblest circumstances. We read about this in 1 Corinthians 1:26-29:

For ye see your calling, brethren, how that not many wise men after the flesh, not many mighty, not many noble, *are called*: But God hath chosen the foolish things of the world to confound the wise; and God hath chosen the weak things of the world to confound the things which are mighty; And base things of the world, and things which are despised, hath God chosen, *yea*, and things which are not, to bring to nought things that are: That no flesh should glory in his presence.

God's glory, revealed in Himself and in His elect, will be apparent to all men and heavenly beings at His coming. At that time, each of God's elect who has died throughout time - along with those of God's children who are alive - will be given a new glorified body designed to last for eternity.

The Unsaved

When we read the Bible, we learn that many people died without ever having been saved, and that there are many people alive today who never will be saved. Of all those who will ever have lived on the earth by the time the Lord puts an end to this universe, there will be many more unsaved than saved persons. Those unsaved persons will never inherit the glorious life that God will provide for the saved. When an unsaved person dies, that's it; there is no longer any hope that the person will ever live again. Imagine that you type something into your computer so that you see it on the screen, but you don't properly save it. Once it disappears from the screen, it's gone - permanently. That will be the situation for an unsaved person who dies. He or she has been lost for all eternity. We read something about this in Ecclesiastes 9:5:

> **For the living know that they shall die: but the dead know not any thing, neither have they any more a reward; for the memory of them is forgotten.**

Many people who realize this as they read the Bible undoubtedly feel like questioning God about this. We must remember, however, that we haven't the right or the wisdom to question God about anything. The Bible tells us about a man named Job. In great suffering, Job felt he had the right to question God's decision to permit his suffering. In Job 23:3-4 we read:

> **Oh that I knew where I might find him!** *that* **I might come** *even* **to his seat! I would order** *my* **cause before him, and fill my mouth with arguments.**

God actually granted Job a reply, but the discussion didn't go as Job thought it would! In Job 38:1-4, we read:

> **Then the LORD answered Job out of the whirlwind, and said, Who** *is* **this that darkeneth counsel by words without knowledge? Gird up now thy loins like a man;**

> **for I will demand of thee, and answer thou me. Where wast thou when I laid the foundations of the earth? declare, if thou hast understanding.**

The Lord continues His reply to Job after this for quite a number of verses, pointing out Job's powerlessness during the creation and his lack of knowledge. Even one of God's children, whom He loves - and we know that God loved Job - has no right to question God about anything. God is so far above man that the wonder is that He would have anything to do with us or that He would choose to save anyone at all.

In Jeremiah 18:6, God compares Himself to a potter who can make a vessel out of clay for any purpose He chooses:

> **O house of Israel, cannot I do with you as this potter? saith the LORD. Behold, as the clay *is* in the potter's hand, so *are* ye in mine hand, O house of Israel.**

Any individual - any "vessel" - may be either saved or unsaved. A commonly held belief in today's local congregations of Christians is that every person who has ever lived has had the opportunity to accept the Lord Jesus as personal Savior. The Bible does not support this position. What we do learn from the Bible is that God uses His word to save people, as in Romans 10:17:

> **So then faith *cometh* by hearing, and hearing by the word of God.**

We also read that we cannot be saved except by the Lord Jesus. In Acts 4:12, the apostle Peter addressed the rulers, elders and scribes of the Jews; speaking of Jesus Christ of Nazareth, he stated:

> **Neither is there salvation in any other: for there is none other name under heaven given among men, whereby we must be saved.**

Sadly, we know that throughout history there have been ages when the populations of entire continents have lived and died without ever hearing the gospel. We must conclude that none of those people were ever saved. This is consistent with the way God dealt with ancient Israel, whenever He instructed them to annihilate a tribe of their enemies - even their women and little ones were put to the sword. Similarly, today each one of us is entirely in God's hands, as we read in Romans 9:15:

> **For he saith to Moses, I will have mercy on whom I will have mercy, and I will have compassion on whom I will have compassion.**

Marriage and Children

God established marriage and the family for mankind, and He has given us many instructions about these relationships. God also uses marriage as a picture of the love He has for those who become His children.

A great deal of misery has resulted because man has not done things God's way. The prevalence of divorce today is an example of this misery. In Mark 10:11-12, we get some insight into God's view of marriage in these words of the Lord Jesus:

> **And he saith unto them, Whosoever shall put away his wife, and marry another, committeth adultery against her. And if a woman shall put away her husband, and be married to another, she committeth adultery.**

Romans 7 also provides scriptures on marriage. We read of God's intended permanence for marriage and the consequence of divorce in God's view in Romans 7:2-3:

> **For the woman which hath an husband is bound by the law to *her* husband so long as he liveth; but if the husband be dead, she is loosed from the law of *her* husband. So then if, while *her* husband liveth, she be married to another man, she shall be called an adulteress: but if her husband be dead, she is free from that law; so that she is no adulteress, though she be married to another man.**

In many places in the Bible, God compares earthly marriage to His relationship with the eternal church consisting of all true believers.

Just as God has promised that His church will be with Him throughout eternity, God has decreed that there is not to be divorce for any reason.

A loving relationship in marriage is the best environment for children, and God's word provides directions on how husband and wife are to live together to create this environment. The husband is the head of the family, but he is to love his wife and be forgiving. In Ephesians 5:28-29, we read:

> **So ought men to love their wives as their own bodies. He that loveth his wife loveth himself. For no man ever yet hated his own flesh; but nourisheth and cherisheth it, even as the Lord the church:**

Also in Ephesians, we read of the man's God-ordained role as head of the family. Ephesians 5:22-24 state:

> **Wives, submit yourselves unto your own husbands, as unto the Lord. For the husband is the head of the wife, even as Christ is the head of the church: and he is the saviour of the body. Therefore as the church is subject unto Christ, so *let* the wives *be* to their own husbands in every thing.**

As head of the family, the man has great responsibility. This includes the responsibility to bring up his children in accordance with God's principles. Many fathers have rejected these principles to the detriment of their own families and communities. In Ephesians 6:4, we read how fathers are to bring up their children:

> **And, ye fathers, provoke not your children to wrath: but bring them up in the nurture and admonition of the Lord.**

Of course, in order to follow this command, fathers must know what the Bible says about bringing up children; and fathers are to teach their children everything that God requires of them. So here too we see a need for diligent Bible study. God emphasizes this in Deuteronomy 6:6-7:

> **And these words, which I command thee this day, shall**

> be in thine heart: And thou shalt teach them diligently unto thy children, and shalt talk of them when thou sittest in thine house, and when thou walkest by the way, and when thou liest down, and when thou risest up.

When we reject the Bible's teaching on any matter, we reject God - just as surely as ancient Israel rejected God from ruling over them so that they could have a king, and just as surely as the the Lord Jesus was rejected by His people. So whenever a decision is made to abort a child, someone has rejected God.

The Bible tells us that God is the author of all life, and that we are not to commit murder; when we read the Bible, we find situations showing us that abortion is always wrong. In Jeremiah 1:5, we read something incredible that God said to the prophet:

> Before I formed thee in the belly, I knew thee; and before thou camest forth out of the womb I sanctified thee, *and* I ordained thee a prophet unto the nations.

Any one of us may eventually be used by God for His own purpose. Who is man to determine that God shall never have the opportunity to use a certain individual, because that individual shall be aborted? Abortion has become a great plague in many nations around the world, and another source of great suffering resulting from our rejection of God and His laws.

The Bible also addresses children and their responsibilities to their parents. In Ephesians 6:1, we read words of God specifically to children:

> Children, obey your parents in the Lord: for this is right.

The words "for this is right" in this scripture seem to be telling us that God has always intended children to be obedient to their parents; it is part of the natural order of things.

Our Operating Manual

If you buy an appliance, power tool, or a complex electronic gadget, you'll almost certainly get a manual from the manufacturer. The manual will provide instructions on how to safely operate the device. Well, the Bible has been called our operating manual. God in His mercy has provided man with an operating manual - about man! It tells us how to live. It tells us what God expects of us, and what He hates. It tells us things about ourselves and our world; some of these are things we could never discover on our own, others are verifiable.

An example of God's advice on how we should live is found in the Book of Philippians. Just as you should be concerned about what goes into your body and should be trying to eat wholesome foods to maintain your physical health, you should also be concerned about what goes into your mind. The Bible provides guidance about this in Philippians 4:8:

> **Finally, brethren, whatsoever things are true, whatsoever things *are* honest, whatsoever things *are* just, whatsoever things *are* pure, whatsoever things *are* lovely, whatsoever things *are* of good report; if *there be* any virtue, and if *there be* any praise, think on these things.**

This certainly sounds like a prescription for mental health - something that is greatly needed in a world of doubt, fear, addiction and suicide. There are many other scriptures telling man how to live. As we learn God's instructions regarding how we are to live, we learn something about God Himself, because we learn what pleases Him. God wants us to do what He tells us to do.

The Bible reveals something interesting about man's life expectancy. Early on, man lived much longer than he now does. Take Noah for example, as we read in Genesis 9:29:

> **And all the days of Noah were nine hundred and fifty years: and he died.**

In the fifth chapter of Genesis there are a number of

Chapter 4 God's Problem

individuals listed with their life spans. All of them were incredibly long lived. No one made it to 1,000 years old, but Methuselah wasn't that far off, as we read in Genesis 5:27:

> **And all the days of Methuselah were nine hundred sixty and nine years: and he died.**

As the millennia passed, man's life expectancy decreased to about what we expect today; this is a number we read of in a prayer of Moses, Psalm 90:10:

> **The days of our years *are* threescore years and ten; and if by reason of strength *they be* fourscore years, yet *is* their strength labour and sorrow; for it is soon cut off, and we fly away.**

Interestingly, the 70 year span (threescore years and ten) prophesied in this Psalm over three thousand years ago agrees well with our life expectancy according to current statistics. It is surely a big undertaking to come up with a number for global life expectancy. Life expectancies vary over a wide range as we go from the industrialized nations (for example, Japan, with men living into the high seventies and women into the low eighties) to the poorest nations of Africa (with life expectancies in the fifties). The World Health Organization, however, has come up with a single number for man's life expectancy. According to a World Health Organization report several years ago, *The World Health 1998*, global life expectancy at birth was 66 years, projected to increase to 73 years by the year 2025. A 70 year life expectancy is roughly what we'll have if we get half the increase predicted by the study, and that could happen roughly half way through the projected time period - in about the year 2011 or so.

Another aspect of our lives covered by the Bible is prayer. God instructs us how we are to pray. In Matthew 6:6, the Lord Jesus taught His disciples about prayer:

> **But thou, when thou prayest, enter into thy closet, and when thou hast shut thy door, pray to thy Father which is in secret; and thy Father which seeth in secret shall reward thee openly.**

The sixth chapter of Matthew also has the verses that have come to be known as the Lord's prayer. This is a prayer in which we ask God for forgiveness, as we read in Matthew 6:12:

And forgive us our debts, as we forgive our debtors.

This is a prayer for salvation. Man needs salvation because God gave man laws by which we are to live, and man has not kept them.

God put laws in the Bible for us to read and study. Many of those laws show us the value that God places on every human life, and cover various situations that may arise to affect a person's life and well-being. There is, for example, a law against kidnapping as we read in Exodus 21:16:

And he that stealeth a man, and selleth him, or if he be found in his hand, he shall surely be put to death.

Of course, the Bible also has a command against murder - as we read in Exodus 20:13:

Thou shalt not kill.

The command against murder must also prohibit suicide, since that is the murder of one's self. There are many today who consider suicide an acceptable solution to certain problems. Some even promote this idea, or teach methods of suicide, or make themselves available to assist someone in committing suicide. Those who do so do not know, or refuse to accept, the Bible's teaching on this matter.

So long as a person is alive, until the Lord Jesus returns to take those who are His elect, there is a chance that God may save that person: even in the last moments of life, and even if the person has lived his entire life trying to ignore God, or believing that Christians are infidels, or believing some other doctrine that is contrary to the truth of the Bible. Once a person has died, however, there is no longer any opportunity for salvation. Suicide is never an acceptable solution.

Adam and Eve had God right there with them in the garden

Chapter 4 God's Problem

of Eden. God could instruct them about life and answer their questions; but we have the complete Bible, everything God wants us to know, and everything we need to know. It's our operating manual.

God's Instructions to Man

What are God's instructions to us? What does the Bible, our operating manual, tell man to do? The Bible tells us that each one of us is under the wrath of God, destined for destruction, unless the Lord Jesus has paid the penalty for our sins. However, it also gives us hope because it tells us that God is merciful. Therefore, we should beg God to be merciful to us, just as the publican - whom we read about in Luke 18:13 -did:

> **And the publican, standing afar off, would not lift up so much as *his* eyes unto heaven, but smote upon his breast, saying, God be merciful to me a sinner.**

We are also to wait on God to save us. It is only by His doing that we can have a heart that is changed so that we want to do His will. God may do this very quickly in our life, or He may do it slowly, over the course of many years. Something the Lord Jesus said helps to clarify this, as we read in John 6:44:

> **No man can come to me, except the Father which hath sent me draw him: and I will raise him up at the last day.**

God draws you to Himself when He saves you. The Bible reveals that God uses His word, the Bible, in this process. Therefore, we should be reading the Bible as much as we can so that we may learn His will. And we should be praying that we might be obedient to His will and be saved.

The Animals

In trying to determine how God sees man, we should also be aware of what the Bible tells us about man's relationship to the animals. From the beginning, we learn that man was given dominion over all the other creatures on the planet. In Genesis 1:27-28, we read:

> **So God created man in his *own* image, in the image of God created he him; male and female created he them. And God blessed them, and God said unto them, Be fruitful, and multiply, and replenish the earth, and subdue it: and have dominion over the fish of the sea, and over the fowl of the air, and over every living thing that moveth upon the earth.**

Apparently, when God made this statement, the dominion He granted to man over all the creatures would have been easier to exercise than it later became. In Genesis 1:30, we read:

> **And to every beast of the earth, and to every fowl of the air, and to every thing that creepeth upon the earth, wherein *there is* life, *I have given* every green herb for meat: and it was so.**

At the beginning, there were no carnivorous animals according to this verse. The nature of many beasts therefore changed some time after they were created. Animals that previously posed no threat to man became dangerous. It could be that this change occurred when God cursed the ground, following Adam's sin. Man's relationship to the animals was very different from then on.

Animals figure in some of the most interesting and important events of the Bible. When God chose to save only Noah and his family out of all mankind in the great flood, He also chose to save specimens of every kind of creature so that they could repopulate the earth. In Genesis 7:7-10 we read:

> **And Noah went in, and his sons, and his wife, and his sons' wives with him, into the ark, because of the waters of the flood. Of clean beasts, and of beasts that *are* not clean, and of fowls, and of every thing that creepeth upon the earth, There went in two and two unto Noah into the ark, the male and the female, as God had commanded Noah. And it came to pass after seven days,**

Chapter 4 God's Problem

that the waters of the flood were upon the earth.

Notice that the animals went in "unto Noah." God selected the creatures He wanted to save, and caused them to go to Noah. Not only that, God must have managed the creatures during the time when they were on the ark. Perhaps He caused them to go into something like a state of hibernation. Here we see a mighty example of God intervening in the animal kingdom to preserve the animals for us, because only those that were on the ark survived.

We read a great deal about animal sacrifice in the Bible, especially in the Old Testament. God called for these sacrifices to show man that payment had to be made for his sins. Rules for animal sacrifices were codified after God called the Israelites out of Egypt, to make a nation of them. (The Israelites are often called the children of Israel in the Bible. They were the descendants of Israel, whose name had been Jacob until God changed it to Israel.)

As we read the Bible, we learn that the sacrifices themselves could never atone for sin for the Israelites; just as nothing we do today, no type or number of good works, can atone for our sins to save us from judgment. The animal sacrifices did, however, point to the sacrifice the Lord Jesus made when He made the payment to satisfy the requirements of God's law on behalf of those God chose to save. In fact, one of the names used for the Lord Jesus is the "Lamb of God," as we read in John 1:29:

> **The next day John seeth Jesus coming unto him, and saith, Behold the Lamb of God, which taketh away the sin of the world.**

In the Old Testament, the institution of the feast called Passover was a powerful picture of the sacrifice that the Lord Jesus, the Lamb of God, would make some 1500 years later. In Exodus 12:3, the Lord is instructing Moses and Aaron about this feast:

> **Speak ye unto all the congregation of Israel, saying, In the tenth *day* of this month they shall take to them every man a lamb, according to the house of *their* fathers, a lamb for an house:**

God's instructions about the lamb for the feast continue on through Exodus 12:6-7, where we read:

> **And ye shall keep it up until the fourteenth day of the same month: and the whole assembly of the congregation of Israel shall kill it in the evening. And they shall take of the blood, and strike *it* on the two side posts and on the upper door post of the houses, wherein they shall eat it.**

And in Exodus 12:12-13, God states why the Israelites must do these things:

> **For I will pass through the land of Egypt this night, and will smite all the firstborn in the land of Egypt, both man and beast; and against all the gods of Egypt I will execute judgment: I *am* the LORD. And the blood shall be to you for a token upon the houses where ye *are*: and when I see the blood, I will pass over you, and the plague shall not be upon you to destroy *you*, when I smite the land of Egypt.**

In these verses about the Passover, we see how God has used the sacrifice of an animal to provide a picture of His salvation plan. Just as the blood of a lamb sprinkled on the door posts protected the occupants of a house when the Lord killed the firstborn in Egypt, the sacrifice of the Lord Jesus - symbolized by His shed blood - will protect the saved from the wrath of God at the end of the world.

Times and Seasons

God's creation operates according to laws He has established, such as the laws of planetary motion. Our way of reckoning time, for example, is determined by the rotation of the earth on its axis every 24 hours; and by the earth's rotation about the sun every 365 days, 5 hours, 48 minutes and 45 seconds; and by the moon's rotation about the earth so that there is a full moon every 29.5 days. These numbers are very consistent, and man has been able to rely on them throughout history.

God also gave us the seven day week at the time of creation, as we read in Genesis 2:1-2:

> **Thus the heavens and the earth were finished, and all the host of them. And on the seventh day God ended his work which he had made; and he rested on the seventh day from all his work which he had made.**

It should be noted that God considers the end of a day to be sundown, which is the beginning of the next day. Interestingly, the number seven doesn't fit perfectly into the monthly cycle, as determined by the time required to go from one full moon to the next. Either a six day week or a five day week would be a better fit: that is, an integral multiple of a week of five or six days would be a closer approximation of the number of days in a month. Yet the seven day week has persisted throughout history. This certainly appears to be an example of God's intervention in the affairs of mankind.

Of course, the earth's movement and position in the solar system affect the cycles on which we all depend, and God has given us His promise of constancy in these matters, as we read in Genesis 8:22:

> **While the earth remaineth, seedtime and harvest, and cold and heat, and summer and winter, and day and night shall not cease.**

Probably, most of us take for granted things as basic as these cycles; yet we are completely dependent on God for them. Happily for us all, God is dependable.

God as Man

The very fact that God, in the person of the Lord Jesus, became man shows that God has a special relationship with man above all other creatures. To give up His glory so that He could live a humble life on earth was in itself a sacrifice that we cannot fathom. Then, when we try to understand the suffering He had waiting for Him at the end of His ministry, and that He knew in advance what He would have to go through, we must stand amazed. And this is all the more incredible for us to contemplate when we realize that He had already paid the full penalty to save His elect before the foundation of the world.

Most of the Gospel accounts deal with the last three years or so of His life. We do know, however, that as a child the Lord Jesus was subject to His parents. This means He was actually subject to His own laws! By obeying His parents, He gave us an example for all time. We read about this in Luke 2:51:

> **And he went down with them, and came to Nazareth and was subject unto them: but his mother kept all these sayings in her heart.**

Even as a human being, the Lord Jesus never ceased to be God; yet He considered those who do God's will to be His family. One day during His ministry, when He was told that His mother and brothers had come and wanted to speak with Him, He used the occasion to tell us something of how He sees those who are obedient to God's law. In Matthew 12:47-50, we read:

> **Then one said unto him, Behold, thy mother and thy brethren stand without, desiring to speak with thee. But he answered and said unto him that told him, Who is my mother? and who are my brethren? And he stretched forth his hand toward his disciples, and said, Behold my mother and my brethren!**

It's startling to think that God would ever think of any man or woman as being like a brother, or sister, or mother to Him; but we have God's word telling us that this is the case. Also, this is the same God who called Abraham His friend, as we read in Isaiah 41:8:

> **But thou, Israel, *art* my servant, Jacob whom I have chosen, the seed of Abraham my friend.**

We also read of Abraham referred to as God's friend in James

2:23:

> And the scripture was fulfilled which saith, Abraham believed God, and it was imputed unto him for righteousness: and he was called the Friend of God.

God was willing to pay a very high price to be able to call any one of us His friend. It meant re-establishing a relationship with man, a relationship that man had broken. It meant making the payment for man's sin, as required by God's law - a law which God Himself follows. This payment carried a high price, and we may never understand how high it was. We know that it was extremely high, because the Gospels give us a glimpse of the Lord Jesus suffering God's wrath in a place called Gethsemane, where He had gone to pray. In Luke 22:44, we read:

> And being in an agony he prayed more earnestly: and his sweat was as it were great drops of blood falling down to the ground.

By going to the cross, the Lord Jesus suffered again and physically died just to demonstrate for us what was required to save us. The fact that He paid this awful price is another indication of God's great love for His people. This is surely one of the most important things we can learn about God.

A little after He had been praying in Gethsemane, the Lord Jesus was taken by a mob led by His disciple Judas Iscariot, who betrayed Him. He was badly beaten, and was crucified the following day. He died that day at just about three in the afternoon, which is called the ninth hour in the Bible. Each of the four Gospels (Matthew, Mark, Luke and John) gives an account of the death of the Lord Jesus. Luke's account has it in chapter 23. In Luke 23:44-46, we read:

> And it was about the sixth hour, and there was a darkness over all the earth until the ninth hour. And the sun was darkened, and the veil of the temple was rent in the midst. And when Jesus had cried with a loud voice, he said, Father, into thy hands I commend my spirit: and having said thus, he gave up the ghost.

The Lord Jesus was crucified and died on a Friday. He was taken down from the cross and buried that evening by one of His disciples, identified in the Bible as Joseph of Arimathea. We read about Joseph's arrangements and the Lord's burial in Mark 15:46-47:

> **And he bought fine linen, and took him down, and wrapped him in the linen, and laid him in a sepulchre which was hewn out of a rock, and rolled a stone unto the door of the sepulchre. And Mary Magdalene and Mary *the mother* of Joses beheld where he was laid.**

These two women came to the tomb early on Sunday morning, roughly a day and a half later, intending to anoint the Lord's body with spices. What they found was an empty tomb, and an angel sitting inside. The angel told them why there was no body in the tomb, as we read in Mark 16:6:

> **And he saith unto them, Be not affrighted: Ye seek Jesus of Nazareth, which was crucified: he is risen; he is not here: behold the place where they laid him.**

Jesus rose from the dead, and was seen by hundreds of His disciples after His resurrection. By His resurrection, the Lord Jesus gave us proof that we may put our trust in Him; and by His resurrection, we have a hope of eternal life that has sustained those whom God has saved throughout all the ages since.

Chapter 5

God's Work

The Bible tells us in many places and in many ways that God works. For example, we have this statement by the Lord Jesus in John 5:17:

> **But Jesus answered them, My Father worketh hitherto, and I work.**

From the very beginning of the Bible, in the Book of Genesis, we read of God's work to create heaven and earth. As we continue reading the Bible, we see God intervening in the affairs of men, and we learn of other work He has done.

Let us consider God's work as being of two general types: the work He has done and continues to do through man, and the work He has done and continues to do apart from man.

God's Future Work as Creator of a New Universe

The creation is, of course, an example of God working apart from man. Man had no active role in the creation whatsoever. Yet the material universe exists to support mankind. God has sustained this universe for a little over 13,000 years now, but He will soon bring it to an end. Mankind will have no say in that matter either, and there is nothing anyone of us can do to stop it. For God's elect, this is good news; for after He destroys this present universe, God will create a new one for His elect to enjoy for all eternity. The Bible mentions this new universe in 2 Peter 3 and elsewhere. In 2 Peter 3:13, we read:

> Nevertheless we, according to his promise, look for new heavens and a new earth, wherein dwelleth righteousness.

The Bible tells us much more about God's work to prepare an individual to inhabit that new universe than it tells about the new universe itself. We will now consider some of the scriptures that tell us about that work.

God's Saving Work

When the Lord Jesus was among us, He revealed the existence of the Father to us. It was the Father who chose those who would be saved, and He made this choice even before the creation of the world. In Ephesians 1:3-6, we read about this:

> **Blessed *be* the God and Father of our Lord Jesus Christ, who hath blessed us with all spiritual blessings in heavenly *places* in Christ: According as he hath chosen us in him before the foundation of the world, that we should be holy and without blame before him in love: Having predestinated us unto the adoption of children by Jesus Christ to himself, according to the good pleasure of his will, To the praise of the glory of his grace, wherein he hath made us accepted in the beloved.**

Notice that the chosen of God are "without blame before him in love," as we have just read in the scriptures. This is accomplished through the atoning work of the Lord Jesus. These scriptures also tell us that it is "according to the good pleasure of his will, To the praise of the glory of his grace, wherein he hath made us accepted in the beloved." We might say that, out of His love for us, God has worked to save us; this in turn brings glory to God.

Of all the billions of people who have ever lived, who are living today, and who will live between now and the end of time, God decided who would be saved. He made His choices even before He created Adam and Eve. God didn't limit His choices to the rich, famous, powerful, or brilliant; although He probably did pick some of these. God tells us that He is no respecter of persons, so most of those He chose are probably people that the world never noticed especially. In 1 Corinthians 1:26-29, we read something about this:

> **For ye see your calling, brethren, how that not many wise men after the flesh, not many mighty, not many noble,** *are called***: But God hath chosen the foolish things of the world to confound the wise; and God hath chosen the weak things of the world to confound the things which are mighty; And base things of the world, and things which are despised, hath God chosen,** *yea*,**and things which are not, to bring to nought things that are: That no flesh should glory in his presence.**

God has often chosen people who are foolish, weak, base or despised by the world. In choosing such people, God is glorified all the more. If God had chosen people who had brilliant minds, or who were powerful and influential, or noble, or adored by the world, then He would be working by our worldly standards. God doesn't need or want our ways of doing things; He works according to His own ways.

Those whom God chose to be saved are no more deserving than those He didn't choose. Why then did God choose them? Only God knows the answer to that - it's God's business. The Bible is clear, however, that we are all sinners and that no one deserves to be saved. In His word, God tells us that He is our Saviour. In Hosea 13:4, we read:

> **Yet I** *am* **the LORD thy God from the land of Egypt, and thou shalt know no god but me: for** *there is* **no saviour beside me.**

In the New Testament, we learn that it is the Lord Jesus who is our Saviour. It was by His work of paying for our sins, at terrible cost to Himself, that He reconciled the elect to God. In Romans 5:10, we read about this reconciliation:

> **For if, when we were enemies, we were reconciled to God by the death of His Son, much more, being**

reconciled, we shall be saved by his life.

It has been taught, possibly without exception until recently, that it was at the cross that the Lord Jesus reconciled us to God by His death. Let us reconsider this idea in the light of some overlooked scriptures. In Revelation 13:8, we read:

And all that dwell upon the earth shall worship him, whose names are not written in the book of life of the Lamb slain from the foundation of the world.

This scripture is telling us that the Lord Jesus, who is often pictured as a lamb in God's word, was "slain from the foundation of the world." There is no doubt that the Lord Jesus was crucified and that He at that time died a terrible death for the sake of the elect; but according to this scripture even before the cross, before the creation of man, the Lord Jesus died for us. We get a further indication that this is so in Hebrews 4:3:

For we which have believed do enter into rest, as he said, As I have sworn in my wrath, if they shall enter into my rest: although the works were finished from the foundation of the world.

There is also a hint of the Lord's death before the Creation in Job 38, where we read the Lord's response to Job. Beginning with Job 38:4, we read:

Where wast thou when I laid the foundations of the earth? declare, if thou hast understanding.

Remember, it was the Lord Jesus by Whom all things were created. Therefore, this verse certainly seems to be identifying the Lord Jesus. The Lord's answer to Job continues, and then in Job 38:17 we read:

Have the gates of death been opened unto thee? or hast thou seen the doors of the shadow of death?

In the question about the gates of death being opened, it sounds like there has been an experience of death. The Lord Jesus knew what it was to die, because that was the price to ransom the elect at

Chapter 5 God's Work

the foundation of the world. Even before the Lord Jesus was born to Mary in Bethlehem in 7 B.C., God was saving people. No one could have been saved unless the Lord Jesus had done the work of paying for that person's sins, and that work was finished long before the Crucifixion.

In the Book of Hebrews, there are many scriptures about someone called Melchisedec. In Hebrews 7:1-3, we read:

> **For this Melchisedec, king of Salem, priest of the most high God, who met Abraham returning from the slaughter of the kings, and blessed him; To whom also Abraham gave a tenth part of all; first being by interpretation King of righteousness, and after that also King of Salem, which is, King of peace; Without father, without mother, without descent, having neither beginning of days, nor end of life; but made like unto the Son of God; abideth a priest continually.**

It is in the Old Testament that we first read about this Melchisedec, who appeared to Abraham many centuries before New Testament times. In fact, this Melchisedec is the Lord Jesus - our high priest. He had already at the time of Abraham become our high priest by offering up Himself, as we read in Hebrews 7:26-27:

> **For such an high priest became us, *who is* holy, harmless, undefiled, separate from sinners, and made higher than the heavens; Who needeth not daily, as those high priests, to offer up sacrifice, first for his own sins, and then for the people's: for this he did once, when he offered up himself.**

The "once" of this scripture must refer to the first time the Lord died, for it was at that time that the payment for our sins was made. What then was the purpose, or the sense, of our Lord dying on the cross? By going to the cross, He demonstrated His love for us. The payment was already made; He didn't have to die for that reason. He died on the cross so that men could know of His love for us and see how great it is.

After dying on the cross, the Lord was immediately in

heaven in His spirit. This is supported by what we read in Luke 23:42-43, in which one of the two thieves crucified with the Lord Jesus asks the Lord to remember him:

> **And he said unto Jesus, Lord, remember me when thou comest into thy kingdom. And Jesus said unto him, Verily I say unto thee, To-day shalt thou be with me in paradise.**

Other scriptures also pertain to this matter. In John 10:17-18, we read:

> **Therefore doth my Father love me, because I lay down my life, that I might take it again. No man taketh it from me, but I lay it down of myself. I have power to lay it down, and I have power to take it again. This commandment have I received of my Father.**

As we read in this scripture, the Lord Jesus told His disciples that He was about to die. He had the power to choose the exact time of His death, and He had the power to resume His physical life. In the tomb, His body didn't begin to decay. This too was prophesied, and was given as a sign for us. The scripture telling us of the Lord's power to take up His life again after the Crucifixion stands in contrast to what we read in Romans 8:11:

> **But if the Spirit of him that raised up Jesus from the dead dwell in you, he that raised up Christ from the dead shall also quicken your mortal bodies by his Spirit that dwelleth in you.**

This scripture indicates that the Lord Jesus was raised up by the Spirit - He didn't raise Himself up. Therefore, this scripture seems to be telling us about the Lord's death at the foundation of the world rather than His death on the cross.

When He died on the cross, the Lord Jesus demonstrated God's love for us. This incredible sacrifice, just to show His children how much He loves them, should make us long to be with Him for eternity. It should erase any fears we may have of what eternal life with God might be like. Perhaps this is what the Lord was telling His disciples in John 14:1-2:

> **Let not your heart be troubled: ye believe in God, believe also in me. In my Father's house are many mansions: if *it were* not *so*, I would have told you. I go to prepare a place for you.**

From another scripture, we know that the Lord did the work of preparing a place for His elect at the foundation of the world; so here, in John 14:2, in telling His disciples that He would "go to prepare a place" for them, He may have been telling them that what He was about to do would change their condition. That is, when the disciples saw how the Lord suffered and died for them on the cross, there would be a new place or condition within their own hearts. When the Lord died at the foundation of the world, He paid the penalty for all of His elect; but when He died on the cross, He demonstrated to all who have ever learned of it - or ever will learn of it - His enormous love for His children.

There is a very interesting scripture that pertains to God's demonstration of His love for the eternal church. In Ephesians 3:10-11, we read:

> **To the intent that now unto the principalities and powers in heavenly *places* might be known by the church the manifold wisdom of God, According to the eternal purpose which he purposed in Christ Jesus our Lord:**

Could the principalities and powers in heavenly places be other types of beings that God created? It appears so, although they won't be discovered by any of our space probes. This scripture also indicates that it was God's intent to make known His wisdom to these beings, by showing them what He has done to save a people for Himself: that is, by creating the eternal church. By His saving work, God demonstrates to us, and to these principalities and powers in the heavens, His mercy, kindness, and especially His love. God has shown us His love in many different ways; but it is certainly the Lord's death on the cross that leaves us in awe of this love.

There is yet more to God's saving work. God uses the hearing of His word to save the elect. When the Holy Spirit applies God's word to a person's heart, that person has been saved: he has received a new soul. In John 3:5-8, we read about this:

> **Jesus answered, Verily, verily, I say unto thee, Except a man be born of water and *of* the Spirit, he cannot enter into the kingdom of God. That which is born of the flesh is flesh; and that which is born of the Spirit is spirit. Marvel not that I said unto thee, Ye must be born again. The wind bloweth where it listeth, and thou hearest the sound thereof, but canst not tell whence it cometh, and whither it goeth: so is every one that is born of the Spirit.**

This incredible work of God, this spiritual rebirth that gives a person a new, resurrected soul, happens when and where God chooses it to happen; but it only happens when a person is under the hearing of God's word and the Holy Spirit is present to save that person because he or she is one of the elect. The apostle Peter witnessed this when he was preaching among the Gentiles, as stated in Acts 11:15:

> **And as I began to speak, the Holy Ghost fell on them, as on us at the beginning.**

The result of this work of the Holy Spirit is that the person has become a new creation, as we read in 2 Corinthians 5:17:

> **Therefore if any man *be* in Christ, *he is* a new creature: old things are passed away; behold, all things are become new.**

The newly saved person has been rescued from being under Satan's power - which is the condition of every unsaved person - and moved into the kingdom of God. The Bible tells us that we should thank God the Father for being rescued in this way. Colossians 1:12-13 states:

> **Giving thanks unto the Father, which hath made us meet to be partakers of the inheritance of the saints in**

light: Who hath delivered us from the power of darkness, and hath translated *us* into the kingdom of his dear Son:

After this has happened, a person still has to contend with his sin-inclined body; but he, in his new life, no longer wants to sin - he wants to do things God's way.

On the day when the Lord calls His elect, each saved person will receive a new, glorified body that will never die. Those of the elect who have died will be resurrected to receive their new bodies, while those elect who are living on that day will be transformed. In 1 Corinthians 15:51-53, we read:

Behold, I shew you a mystery; We shall not all sleep, but we shall all be changed, In a moment, in the twinkling of an eye, at the last trump: for the trumpet shall sound, and the dead shall be raised incorruptible, and we shall be changed. For this corruptible must put on incorrupttion, and this mortal *must* put on immortality.

For the elect, this will complete God's saving work.

Everyone Fits Into One of These Categories

As we reflect on what the Bible teaches us about God's saving work, we see that there are two major categories of people. These categories apply to all people who have ever lived or ever will live: the elect (those God decided to save), and the non-elect (those God decided not to save). We learned that God made His decision at the foundation of the world. We also learned that God's decision is being implemented through His salvation plan on a large scale in major events throughout history (e.g., the great flood, the birth of the Lord Jesus, the end of the church age) and on a small scale as He works in the lives of individuals. God's work in the lives of those individuals He chose to save leads us to deduce three subcategories for these people, who are the elect:

The Elect

1. Those who have died and are now with the Lord.
2. Those who are presently alive and saved.
3. Those who are presently alive, although not yet saved.

These subcategories have been applicable throughout most of human history - beginning at or shortly after the death of Adam and Eve's son, Abel, and continuing right up to this day. These subcategories will continue until May 21, 2011. By that day, God will have completed His saving work. Anyone who is now in that third subcategory will have been saved by then. Until then, God is committed to saving them. The Lord Jesus has already paid for their sins, and God is waiting for the right time in each person's life to give that person a new, resurrected soul. By the day of the rapture, each of the elect who is alive will have received that new resurrected soul; they will join all the elect who have died, and each one - in an immortal body - will be caught up in the air to meet the Lord.

We may also deduce subcategories for the non-elect:

The Non-Elect

1. The non-elect who have died or will have died by May 21, 2011.
2. The non-elect who will be alive on May 21, 2011.

The non-elect who have died before judgment day begins will not suffer any more. They will never live again. They will be spared the terror of judgment day.

The non-elect who remain alive on judgment day will experience God's wrath.

They may die on that day as a result of the events beginning then, or they may die soon afterwards. There will be no hope for these people, only degrees of suffering.

God Intervenes

We know that in the last days God will save a great multitude. We also know - at least there is every indication of this - that before these last days God did not save very many people.

There was a "harvest" of God's elect during the church age, but this was not the great harvest we read about in the Bible; and judging from what we do read there, during all the thousands of years before the church age began, relatively few persons were saved. There is a startling implication in all of this. In order to save someone in the last days, God may have had to intervene in the lives of hundreds of his ancestors, even if none of them were saved.

Let's suppose God wanted to save a man born in the 1990's, and that none of his current family or his ancestors from thousands of years ago had been saved. In order to bring that person to the time and place where God could save him in our day, God may have generously blessed many of his ancestors. God would have had to arrange things so that this man's ancestors survived and bore the next generation. Possibly, God saved many of their lives and intervened many times in their affairs. God may, for instance, have awakened one of them at just the right moment in order to save him from a wild animal that was about to pounce on him; or perhaps God caused it to rain on a small piece of land just to save a crop that otherwise would have been lost, and so saved a family from starvation; or maybe God arranged for a girl to be in the right place at the right time to meet the boy she would eventually marry. God may have intervened in such ways on many occasions to preserve the lives of those people - even though they were not God's elect - in order to allow just one of His elect to be born at a certain time in history and in a certain place, and to live his life in a way that led to the circumstances in which he would be saved. This is indicated in Acts 17:26-27:

> **And hath made of one blood all nations of men for to dwell on all the face of the earth, and hath determined the times before appointed, and the bounds of their habitation; That they should seek the Lord, if haply they might feel after him, and find him, though he be not far from every one of us:**

All of this represents an enormous amount of work on God's part. It's as if God has been watching over each one of His elect ever since creation.

In the Bible, we find many situations in which God intervened in the lives of individuals and nations. For example, the kingdom of Judah faced a great army during the reign of king Asa, as we read in 2 Chronicles 14:9-12:

> **And there came out against them Zerah the Ethiopian with an host of a thousand thousand, and three hundred chariots; and came unto Mareshah. Then Asa went out against him, and they set the battle in array in the valley of Zephathah at Mareshah. And Asa cried unto the LORD his God, and said, LORD, *it is* nothing with thee to help, whether with many, or with them that have no power: help us, O LORD our God; for we rest on thee, and in thy name we go against this multitude. O LORD, thou *art* our God; let not man prevail against thee. So the LORD smote the Ethiopians before Asa, and before Judah; and the Ethiopians fled.**

Of course, there will be many who read this and say that Asa also had a very substantial army - which he did - and that Judah just had a better army than the Ethiopians. The Bible, however, clearly tells us that it was God who was responsible for Judah's victory.

Perhaps the most familiar instances of God working to intervene in the lives of individuals are some of the miracles performed by the Lord Jesus. In Matthew 20:31-34, for instance, we read about two persistent blind men who called out to the Lord Jesus as He passed by them:

Chapter 5 God's Work

And the multitude rebuked them, because they should hold their peace: but they cried the more, saying, Have mercy on us, O Lord, *thou* son of David. And Jesus stood still, and called them, and said, What will ye that I shall do unto you? They say unto him, Lord, that our eyes may be opened. So Jesus had compassion *on them*, and touched their eyes: and immediately their eyes received sight, and they followed him.

In the Gospels, we read the Lord's plain statement to His disciples that He spoke in parables. We might say that a parable is an earthly story with a heavenly meaning. They are simple stories, but they contain spiritual truths that may be difficult to understand. The story about the Good Samaritan and the one about the Prodigal Son are well known examples of these parables. Although the Lord's parables may be familiar, it is not commonly recognized that it is not only the parables that contain spiritual truths - we need to look for spiritual truth in every situation in the Bible. Any situation could be a tableau or picture illustrating some aspect of God's salvation plan.

Whenever we read of a conversation, or a miracle, or any occurrence in the Bible, we may be certain that it actually took placed as described; but we should realize that God has planned all of these things to happen exactly as they did. Whenever we discover spiritual truth in one of these events, we not only come to a deeper understanding of things God is teaching throughout the Bible, we also see proof that the Bible is the work of God.

The account of the two blind men who received their sight as a result of the Lord's miraculous healing is like a parable in action. It was an actual event, but it shows us something about salvation. There is every indication that the Lord saved those two men that day the instant they received their sight. It happened immediately, just as life entered immediately into those whom the Lord raised from the dead. In an instant, these two men each received a new, resurrected soul; and they were changed. They followed the Lord, because after someone is saved, that person wants to do the Lord's will. God's intervention in the lives of those two men that day did much more than provide them with the ability to see. They were now born of the Spirit, and so had eternal life.

God's Work to Create and Preserve His Word

God surely planned out the Bible just as He planned in detail how He would save the elect before the foundation of the world. His planning for the Bible would have included working in the affairs of man to develop the very languages that He used to write the Bible.

There was a time when all mankind shared a common language. In Genesis 11:1, we read:

> **And the whole earth was of one language, and of one speech.**

That time ended when God foiled the plans of some men to build a city with a great tower. Perhaps these men dreamed of riches, and realized they needed a great population center to achieve their goals. The city and the famous tower would induce people to remain in the area and bring in others. In Genesis 11:4, we read about their goal:

> **And they said, Go to, let us build us a city and a tower, whose top *may reach* unto heaven; and let us make a name, lest we be scattered abroad upon the face of the whole earth.**

The plans of these men, however, conflicted with God's plans for mankind; and so, in Genesis 11:7-9, we read what the Lord decided to do:

> **Go to, let us go down, and there confound their language, that they may not understand one another's speech. So the LORD scattered them abroad from thence upon the face of all the earth: and they left off to build the city. Therefore is the name of it called Babel; because the LORD did there confound the language of**

all the earth: and from thence did the LORD scatter them abroad upon the face of all the earth.

We know that God's plans are incredibly complex, involving untold billions of details. Most likely, from the day He confused the language at Babel, God began to guide the development of the Hebrew and Greek languages in which the Bible was originally written. God could have known beforehand every word, every letter, he wanted to use; so He would have seen to it that these languages developed as He wanted them.

God must also have been guiding the lives of the many people whose names we read in the Bible. God certainly knew beforehand the men whom He would use to write down His words to create the 66 books that comprise the Bible; but He also knew - from the foundation of the earth - all the persons whose stories we read there, including those who are mentioned only by name so that we know nothing more about them. God knew them all, and knew how He would intervene in their lives to create the Bible. God knew what they would do, what they would say, and where they would be; and God knew what He would write about them. We can't understand how God can do these things, but the Bible tells us He can. In Isaiah 46:9-11, we read:

> **Remember the former things of old: for I *am* God, and *there is* none else; *I am* God, and *there is* none like me, Declaring the end from the beginning, and from ancient times *the things* that are not *yet* done, saying, My counsel shall stand, and I will do all my pleasure: Calling a ravenous bird from the east, the man that executeth my counsel from a far country: yea, I have spoken *it*, I will also bring it to pass; I have purposed *it*, I will also do it.**

God finished writing the Bible about 65 years or so after the Crucifixion, but His work to preserve His word continued. Unless it had, we wouldn't have the Bible today. There have been many attempts to destroy the word of God throughout history. Satan has probably been responsible for most of these attempts, if not all of them. Until the invention of the printing press - and even for quite some time afterwards - copies of the Bible were scarce. The Bible could easily have been lost to posterity, as undoubtedly many of the works of antiquity have been. God did not let that happen. In Mark 13:31, we read something that the Lord Jesus said about this:

Heaven and earth shall pass away: but my words shall not pass away.

The ancient kingdom of Judah also had this promise from God, as we read in Isaiah 40:8:

The grass withereth, the flower fadeth: but the word of our God shall stand for ever.

God uses His word to save people, and so the Bible had to survive to our day; because it is in our day that God is saving so many people. After the invention of the printing press, Bibles eventually became commonplace.

In our period of history, there have been incredible technological advances. It has only been in the last two hundred years or so that the major scientific discoveries that have shaped our lives occurred. God has surely been behind this explosion in scientific knowledge. During the church age, Bibles were printed up by the tens of thousands and distributed all over the planet. Still there are locations where - for a variety of reasons - it is not physically possible to reach people with a Bible. In our day however, technology has permitted barriers to the printed word to be overcome. This is especially true of radio. It may be illegal to bring Bibles into a certain country, or to stand on a street corner reading the Bible to passersby; but radio broadcasts aren't stopped by such government policies. In this way, many millions of people who can't have a Bible can hear God's word.

When we read the Bible, we find that God is concerned about His glory. The creation of the Bible is proof of His glory, for only God could have created it. Only He could have written it in such a way; for so many things that were hidden until recently were right there, under our noses all the time. Yet we couldn't see those things until the time He chose to reveal them.

Chapter 5 God's Work

God Equips Us To Do His Will and Works Through Us

In Ephesians 2:10, there's an amazing scripture that applies to God's elect; there we read:

For we are his workmanship, created in Christ Jesus unto good works, which God hath before ordained that we should walk in them.

This scripture indicates that God has ordained or prepared good works for His people to perform. Since God knew each of His elect even before the creation, it follows that God must have prepared these people for the works He planned for them to do. He must have equipped them with the natural talents and abilities they would need, and He must have caused them to live under such conditions that they would eventually have the skills and other resources needed to carry out their assigned tasks.

What kind of tasks might these be? For one thing, the Lord's command to preach the Gospel applies to anyone who has been saved. In Mark 16:15, we read this command of the Lord Jesus, given to His disciples after He had risen from the grave:

And he said unto them, Go ye into all the world, and preach the gospel to every creature.

Of course, it is not possible for every one of God's elect to pack up and head for distant lands as a missionary. Probably, God has prepared relatively few to do that. No matter where we are, however, we may still carry out the Lord's commission. We may do this by telling those with whom we regularly come in contact, at an opportune time, about the God of the Bible. We may also distribute Gospel tracts, or contribute financially to the work of getting the Gospel out to the world.

Each of God's elect should be constantly aware that he is to live for the Lord every day of the week. There is, however, a special day for God's people in this New Testament era: Sunday. The seventh day Sabbath of the Old Testament has been replaced by a new Sabbath, which is the Sunday Sabbath. Beginning at sundown on Saturday - for that is when the new day begins - until the sun goes down late on Sunday, God wants us to put aside our usual activities. Most Christians today will be found attending their local congregation on a Sunday. We shall see conclusive proof, however, that God has finished using the local congregations: the church age has ended. How, then, does God want Sunday to be observed. Unlike the Saturday Sabbath of the Old Testament, which is no longer in force, the Sunday Sabbath is not a day of rest. It is a day to get the Gospel out, if we have a means to do so. It is also a day for us to read and meditate on God's word, and to pray.

Even though it is not required that Sunday be observed as a day of rest, we must not assume that it's acceptable to do all kinds of work on that day. Obviously, some occupations and professions require work on Sunday. If you are seriously injured on a Sunday and need to go to a hospital, you'll certainly want a doctor to take care of you when you get there. Firemen, policemen, doctors, nurses and many others who provide essential and emergency services are required to work on Sundays. The rest of us, however, should try to arrange our work from Monday through Saturday so that we may keep Sunday available for spiritual things.

God wants us to set a good example to others every day of the week. As the Lord Jesus instructed His disciples in Matthew 5:16:

> **Let your light so shine before men, that they may see your good works, and glorify your Father which is in heaven.**

It should eventually become our nature to behave in such a way that we do set that good example. This is so because a person is changed and set on a new path in life after he has received his new, resurrected soul. That path will lead to further change. In Galatians 5:22-23, we read of the traits that a saved person should be developing:

> **But the fruit of the Spirit is love, joy, peace, longsuffering, gentleness, goodness, faith, Meekness, temperance: against such there is no law.**

These traits permit the world to see the light in the lives of God's children, which is the reflection of the Lord Jesus.

The command to preach the Gospel was given in the New Testament, and so it did not apply to ancient Israel. Of course, God was using individuals to accomplish His work at that time too; and He was equipping them as required to do whatever had to be done. The life of Moses is surely an illustration of this. He was raised by Egyptian royalty, and he had the advantage of being educated in what was probably the most advanced civilization of that time. Consequently, he was uniquely qualified to deal with Pharaoh and his court when the time came.

There is another example of how God equips individuals for their tasks, also from the time of the Exodus. When the children of Israel were in the wilderness, God instructed them to build the tabernacle. God gave to Moses plans for the tabernacle and items that were to be made and placed within it. These plans are recorded in the Book of Exodus. The tabernacle was actually a great tent where the Lord created a visible sign of His presence. In Exodus 40:34, we read what happened when the tabernacle was finished:

Then a cloud covered the tent of the congregation, and the glory of the LORD filled the tabernacle.

The tabernacle, including all the items that were to be made and placed within it, was to be a thing of beauty. It would require great skill to work with the different types of materials and to execute the work in the detail required by the plans. In Exodus 31:1-6, we read the names of men whom God called to do this work:

And the LORD spake unto Moses, saying, See, I have called by name Bezaleel the son of Uri, the son of Hur, of the tribe of Judah: And I have filled him with the spirit of God, in wisdom, and in understanding, and in knowledge, and in all manner of workmanship, To devise cunning works, to work in gold, and in silver, and in brass, And in cutting of stones, to set *them*, and in carving of timber, to work in all manner of workmanship. And I, behold, I have given with him

Aholiab, the son of Ahisamach, of the tribe of Dan: and in the hearts of all that are wise hearted I have put wisdom, that they may make all that I have commanded thee;

Clearly, God worked in the lives of those men to equip them to build the tabernacle. He blessed them with the skills and intelligence needed to perform the wide variety of tasks associated with such a project.

We must not assume, however, that God has saved an individual just because He has blessed him in some special way. Alexander the Great, for example, was blessed by God with the tremendous abilities that allowed him to conquer so much of the world. Alexander was certainly used in God's plans, for it was as a result of his conquest that the Greek language became widely used. As we have seen, God chose to write the New Testament in Greek. Alexander was certainly serving God's purpose as he waged war for his own glory. Yet Alexander was definitely not one of God's elect. In fact, the Bible uses Alexander as a picture of Satan. Therefore, we see that God may equip anyone - saved or unsaved - to do whatever His plan requires.

God's Intervention Affects Satan

The Book of Job tells us something surprising about Satan. It shows that there was once a time when Satan had access to heaven, to actually come before God and speak with Him; and this was after the time when Satan had shown himself to be evil! In Job 1:6-7, we read:

> **Now there was a day when the sons of God came to present themselves before the LORD, and Satan came also among them. And the LORD said unto Satan, Whence comest thou? Then Satan answered the LORD, and said, From going to and fro in the earth, and from walking up and down in it.**

As we read this amazing story, we learn that God - for His own purposes - allowed Satan to bring terrible suffering to Job. God's work of intervention is not, however, limited to the affairs of mankind. The Bible indicates that Satan's power was diminished after the Crucifixion.

Examples of Satan's earlier power surely are the miracles that Pharaoh's magicians performed when they were confronted by Moses and Aaron. The miracles God worked through Moses and Aaron were far greater than those worked by the magicians. Yet, it is astonishing that the magicians could have worked any miracles at all. An example of the magicians' miracles is recorded in Exodus 7:10-12:

> **And Moses and Aaron went in unto Pharaoh, and they did so as the LORD had commanded: and Aaron cast down his rod before Pharaoh, and before his servants, and it became a serpent. Then Pharaoh also called the wise men and the sorcerers: now the magicians of Egypt, they also did in like manner with their enchantments. For they cast down every man his rod, and they became serpents: but Aaron's rod swallowed up their rods.**

Satan's freedom to work as he did through those magicians apparently continued until the time the Lord Jesus was crucified. Until then, his power was great. In fact, even as the hour when the Lord would be arrested drew near, Satan himself entered into Judas Iscariot - one of the Lord's twelve apostles. The Lord handed a morsel of food to Judas at the Passover meal they were sharing, and that is when Satan entered Judas. We read about this in John 13:27:

> **And after the sop Satan entered into him. Then said Jesus unto him, That thou doest, do quickly.**

Apparently, only the Lord knew what had happened. The other eleven apostles didn't seem to notice any change in Judas. A short time later, Judas betrayed the Lord to the chief priests and Pharisees.

After the Crucifixion, Satan's situation changed. The Lord

had spoken about this before His death, but His disciples may not have understood. In Luke 11:21-22, we read a very interesting statement He made:

> **When a strong man armed keepeth his palace, his goods are in peace: But when a stronger than he shall come upon him, and overcome him, he taketh from him all his armour wherein he trusted, and divideth his spoils.**

The "strong man" was overcome at the cross. Satan was bound at that time, as we read in Revelation 20:1-2:

> **And I saw an angel come down from heaven, having the key of the bottomless pit and a great chain in his hand. And he laid hold on the dragon, that old serpent, which is the Devil, and Satan, and bound him a thousand years,**

After the cross, Satan had less power; but God didn't leave him powerless. In fact, throughout the church age, Satan was sowing tares - weeds - throughout the churches. At the beginning of the church age, God poured out the Holy Spirit and began the next phase of His salvation plan. We are all in Satan's kingdom until God saves us. When God saves someone, he plucks that person out of Satan's kingdom and so "divides his spoils." God kept Satan in this restricted condition throughout the church age; but when the church age ended, God actually set him up to rule in the churches.

Although the language of the Book of Revelation is seemingly impossible to understand, many of its scriptures begin to make sense once God's time-line is understood. We now understand that the time of Satan's rule over the churches (the local congregations) is reflected in Revelation 20:7-8:

> **And when the thousand years are expired, Satan shall be loosed out of his prison, And shall go out to deceive the nations which are in the four quarters of the earth, Gog and Magog, to gather them together to battle: the number of whom *is* as the sand of the sea.**

God had "bound" Satan, and then God "loosed" him "out of his prison." Although Satan is very powerful, he can't do anything unless God allows it.

Knowing God's time-line, we can know that the figure of a thousand years that we read about in Revelation 20 is to be understood figuratively. It represents completeness of the time when God restrained Satan and worked through the local congregations. Now, in this relatively brief period between the end of the church age and the end of the world, Satan has apparently regained at least some of his earlier strength.

The Bible indicates that God is no longer using spectacular signs and wonders. We know that for many people, the Lord will come "as a thief in the night." For this reason, we know that the spectacular events in the Book of Revelation cannot be interpreted literally. The events must be understood spiritually. We see an indication of this in something the Lord Jesus said to the Pharisees; in Luke 17:20, we read:

> **And when he was demanded of the Pharisees, when the kingdom of God should come, he answered them and said, The kingdom of God cometh not with observation:**

Satan, on the other hand, has regained strength so that he can work miracles as he uses the churches to bring a false gospel to the world. The Lord Jesus warns us about this in Matthew 24:23-24:

> **Then if any man shall say unto you, Lo, here *is* Christ, or there; believe *it* not. For there shall arise false Christs, and false prophets, and shall shew great signs and wonders; insomuch that, if *it were* possible, they shall deceive the very elect.**

Perhaps, in Satan's warped mind, he still thinks he can win. The Bible tells us that Satan is a liar and the father of lies. Maybe he has deceived himself into thinking he can one day rule the universe; but the Bible tells us that his fate will be the same as that of all the unsaved people of the world: Satan and all his demons will be destroyed on the earth's last day.

Chapter 6

God's Calendar

There is something about the Bible that is really amazing. It's something that many people have never heard before. It is this: the Bible contains a calendar. This means that by analyzing the Bible we can determine the dates of occurrence for many of the major events written about in the Bible. How, you may be wondering, is this possible?

In order to understand the Biblical calendar, we first need to realize that God set in motion a celestial time keeping system when He created the universe. Ever since then, a day has always been the same - barring an occasion of Divine intervention - because the length of a day is determined by the spin of the earth about its axis. Likewise, the length of a year has always been the same because it is determined by the rotation of the earth about the sun.

There can be confusion when an attempt is made to use months to determine the passage of time between Biblical events. At the time of Noah, a month consisted of 30 days. We may see this by comparing scripture with scripture using Genesis chapters 7 and 8. In Genesis 7:11, we read:

> **In the six hundredth year of Noah's life, in the second month, the seventeenth day of the month, the same day were all the fountains of the great deep broken up, and the windows of heaven were opened.**

And in Genesis 8:3-4, we read:

And the waters returned from off the earth continually: and after the end of the hundred and fifty days the waters were abated. And the ark rested in the seventh month, on the seventeenth day of the month, upon the mountains of Ararat.

When we see how much time elapsed between Genesis 7:11 and Genesis 8:4, we realize that it was exactly five months; but Genesis 8:3 tells us this period of time was 150 days. Therefore, we have an indication that there were 30 days to a month at that time in history.

Just before the Lord brought Israel out of Egypt, when He instituted the Passover, He gave them a new calendar. We read this in Exodus 12:1-2:

And the LORD spake unto Moses and Aaron in the land of Egypt, saying, This month *shall be* unto you the beginning of months: it *shall be* the first month of the year to you.

That calendar may have been very much like the lunar calendar used by many people today for religious observances. A lunar calendar relies on the cycle of the moon about the earth, averaging about 29.5 days. On the other hand, in our modern calendar the number of days in a month may be as few as 28 or as many as 31.

When we use years and days to determine the passage of time, we may avoid any complications of switching between different types of months. For example, if we say an event happened in 4990 B.C., we mean the number of years that has passed since that event equals 4,990 plus the number assigned to the current year (A.D.) minus one (because there is no year "0").

The need to subtract one when determining the passage of time from a B.C. date to an A.D. date is a frequent source of confusion. Imagine that you are living in the year 1 B.C. and that you continue living in this time period for one year. The year that follows the year we call 1 B.C. is the year we call 1 A.D. - there is no year in between; so after that one year has gone by, you are living in the year 1 A.D. You have seen only one year go by, but mathematically (because our calendar assigns the numbers 1 B.C. and 1 A.D. respectively to those two years) it looks like two. That's because, mathematically, the difference between a negative (-) 1 and a positive (+) 1 is 2; but when we are counting the years between 1 B.C. and 1 A.D., there's only one. As we consider the Biblical calendar, we need to keep this in mind.

Significance of Numbers in the Bible

We also need to look at the way God uses numbers in the Bible before we consider the Biblical calendar. You really can't talk about time without using numbers, and God has placed many numbers in the Bible. God uses certain numbers in ways that allow us to associate spiritual meanings with those numbers. Let's take a look at a use of the number two in the Bible. We see it used in Luke 10:1:

> **After these things the Lord appointed other seventy also, and sent them two and two before his face into every city and place, whither he himself would come.**

From this scripture and other appearances of the number two in the New Testament, we may say that it is associated with those who bring the Gospel.

We may not see a spiritual significance for a given number in every scripture where it appears. Perhaps there are scriptures for which the numbers therein have no spiritual significance; or, if the spiritual significance is there, perhaps God hasn't revealed it to anyone yet. This may be the case with the number two, which appears many times in the Old Testament. There, it doesn't seem to be associated with the spiritual meaning it appears to have in the New Testament. When we see how it is used in Luke 10:1 and elsewhere in the New Testament, we are led to conclude that God is associating it with those who bring the Gospel.

Continuing with our examination of numbers in the Bible, let's take a look at the number three. For an example of the use of the number three, read 1 Chronicles 21:9-10:

> **And the LORD spake unto Gad, David's seer, saying, Go and tell David, saying, Thus saith the LORD, I offer thee three** *things*: **choose thee one of them, that I may do** *it* **unto thee.**

If we continue reading this chapter of 1 Chronicles, we find that each of the three choices is terrible (famine, pestilence, or war). In the Gospel accounts of the Crucifixion, we also see appearances of people and things in threes. For example, from John 19:18-20, we know that there were three crosses and that there was writing on the Lord's cross in three languages:

> **Where they crucified him, and two other with him, on either side one, and Jesus in the midst. And Pilate wrote a title, and put** *it* **on the cross. And the writing was, JESUS OF NAZARETH THE KING OF THE JEWS. This title then read many of the Jews: for the place where Jesus was crucified was nigh to the city: and it was written in Hebrew,** *and* **Greek,** *and* **Latin.**

Just as it was God's purpose during David's reign to punish Israel in one of three ways, it was also God's purpose that the Lord Jesus should go to the cross. We may say that, spiritually, the number three signifies God's purpose.

Let's continue with more examples to see how some other numbers are used: for example, the number five. In Matthew 25:1-2, we read:

> **Then shall the kingdom of heaven be likened unto ten virgins, which took their lamps, and went forth to meet the bridegroom. And five of them were wise, and five** *were* **foolish.**

As we read the parable, we find that the five who were wise were saved; the five who were foolish were not. We may say that the spiritual significance of the number five is salvation through the atonement of the Lord Jesus, or judgment.

The number seven is used throughout the Bible to indicate a

Chapter 6 God's Calendar

sense of completeness. It's used in this way by the Lord in His reply to the apostle Peter after he had asked how many times he was to forgive someone who sinned against him. We read in Matthew 18:21-22:

> **Then came Peter to him, and said, Lord, how oft shall my brother sin against me, and I forgive him? till seven times? Jesus saith unto him, I say not unto thee, Until seven times: but, Until seventy times seven.**

The Lord's use of multiplication in this scripture is very interesting. Could this be a clue about how we are to deal with certain numbers? In fact, we will see that the Bible frequently gives us numbers that may be broken down - that is, factored - into the smaller numbers of spiritual significance that we are currently considering.

The number ten, and any number you can get by multiplying it by itself (such as 100, 1000, and 10,000), appears to signify completeness. We see this in Luke 15:4, in which the Lord Jesus is comparing His elect to sheep:

> **What man of you, having an hundred sheep, if he lose one of them, doth not leave the ninety and nine in the wilderness, and go after that which is lost, until he find it?**

Another number of great interest is the number 17. In the Book of Jeremiah, we find that Jeremiah paid 17 shekels to buy a piece of land in Judah, shortly before Judah was to be conquered. God had revealed to Jeremiah that the king of Babylon would indeed conquer Judah, and He told Jeremiah to buy the land despite this. In Jeremiah 32:9, we read:

> **And I bought the field of Hanameel my uncle's son, that *was* in Anathoth, and weighed him the money, *even* seventeen shekels of silver.**

Jeremiah's purchase of the field was to be a sign that the people of Judah would one day return. In fact, this did happen 48 years after the fall of Judah. The spiritual meaning of Jeremiah's purchase of the field, however, appears to be that God will some day bring His people to heaven. This is indicated in Jeremiah 32:41-42, in which God is talking about His people:

> **Yea, I will rejoice over them to do them good, and I will plant them in this land assuredly with my whole heart and with my whole soul. For thus saith the LORD; Like as I have brought all this great evil upon this people, so will I bring upon them all the good that I have promised them.**

Spiritually, then, the number 17 appears to signify heaven.

The number 23, indicating God's wrath or judgment, is also of great interest. We see it used in Daniel 8:14, where the prophet Daniel has recorded something he saw and heard in a vision:

> **And he said unto me, Unto two thousand and three hundred days; then shall the sanctuary be cleansed.**

When we realize that the number in this scripture (2,300) is equal to 23 x 100, and we see how it is used in the context of Daniel chapter 8, we may see this scripture as an illustration of the spiritual meaning of the number 23. The number 100 indicates the completeness of whatever is in view, which in this case seems to be God's wrath or judgment (the number 23).

We find that the number 37 also signifies God's wrath or judgment. In Genesis chapters 7 and 8, we see an example of the use of this number. From Genesis chapter 7, we know that Noah was 600 years old when the flood occurred. We also learn that it was in the second month, on the seventeenth day that "all the fountains of the great deep were broken up, and the windows of heaven were opened." Compare this with Genesis 8:14, where we read:

> **And in the second month, on the seven and twentieth day of the month, was the earth dried.**

This scripture refers to Noah's six hundred and first year. Therefore, we see that the earth was dried one year and ten days after that day the flood began. This works out to 370, or 37 x 10, days. Therefore, like the number 23, the spiritual significance of the number 37 seems to be God's wrath or judgment (with the number 10, as a factor of the number 370, indicating completeness).

Another number that appears to have spiritual significance is the number 43. In the Book of Exodus, we find that the children of Israel were in the land of Egypt for a total of 430 years. We see the number 43 when we factor 430 to get 43 x 10, in which the number 43 appears to indicate God's wrath or judgment (the number 10 again indicates completeness).

The numbers 4, 12 and 40 also appear throughout the Bible. When we consider these numbers - and this also applies to the number 10 - we do not try to break them down into smaller numbers of spiritual significance. Rather, we consider them as they are in the scriptures. Let's first look at the number 4.

As an example of how the number 4 is used, read Psalm 107:2-3:

Let the redeemed of the LORD say *so*, whom he hath redeemed from the hand of the enemy; And gathered them out of the lands, from the east, and from the west, from the north, and from the south.

From this scripture and others in the Bible, we take the number 4 (as in the four points of the compass) to have a spiritual significance meaning the farthest extent in time or in distance that God spiritually has in view.

The number 12 is seen often in the Bible. Many people know about the 12 apostles and the 12 tribes of Israel. The spiritual significance of the number 12 appears to be the fullness of whatever is in view.

The number 40 also appears often. In Matthew 4:1-2, we see the use of this number in connection with the Lord Jesus:

Then was Jesus led up of the Spirit into the wilderness to be tempted of the devil. And when he had fasted forty days and forty nights, he was afterward an hungered.

In this scripture and others, the number 40 has a spiritual significance of testing.

The numbers 11 and 13 also have spiritual significance. For all the other numbers of spiritual significance, we saw meaning for each number in one or more scriptures. For the numbers 11 and 13, however, we will see spiritual significance from the Biblical calendar itself. Later on, when we examine the calendar, we will see how the number 11 is associated with the first coming of the Lord Jesus and how the number 13 is associated with the end of the world.

The spiritual significance of numbers in the Bible presents a strong argument against the use of any translation of the Bible that converts Biblical units of measurement, thereby changing the original numbers. For example, the cubit is a unit of length we come across frequently when we read the Bible. It is believed to be equal to about 18 inches. Genesis 6:15, in which God is telling Noah how big the ark shall be, gives the dimensions in cubits as follows:

And this *is the fashion* which thou shalt make it *of*: The length of the ark *shall be* three hundred cubits, the breadth of it fifty cubits, and the height of it thirty cubits.

From this verse, we see that the ark's dimensions may be listed as follows:

Length:	3 x 100
Breadth	5 x 10
Height	3 x 10

We see in these dimensions three numbers (three, five and ten) that have been identified as having a specific spiritual significance. Using the meanings previously identified for these numbers, we see how the ark's very dimensions seem to be telling us that it is God's complete (10) purpose (3) to apply the atonement (5), thereby bringing salvation and judgment. The thought of the Lord's atonement for our sins, in connection with a flood that destroyed almost all of mankind, may seem incongruous. Remember, however, that the ark was a giant lifeboat for those whom God chose to save. For Noah and his family, there was salvation; for everyone else, there was judgment.

We may dig even deeper into the verse giving us the ark's dimensions, looking for more spiritual meaning. The words used for length, breadth and height are used elsewhere in the Bible in ways that give us a clue that God is associating those words with other meanings. For example, in Psalm 21:4, we read:

> **He asked life of thee, *and* thou gavest *it* him, *even* length of days for ever and ever.**

The same Hebrew word (Strong's number 753) used for "length" when giving the ark's dimensions is also used here and is linked to the thought of time ("for ever and ever"). Similarly, the word used for "breadth" in the ark's dimensions (Strong's number 7341) is used in the last part of Isaiah 8:8:

> **and the stretching out of his wings shall fill the breadth of thy land, O Immanuel.**

The phrase "breadth of thy land" seems to suggest all the world. Finally, the word used for "height" (Strong's number 6967) in connection with the ark is also found in the original Hebrew of Ezekiel 31:10-11:

> **Therefore thus saith the Lord GOD; Because thou hast lifted up thyself in height, and he hath shot up his top among the thick boughs, and his heart is lifted up in his height; I have therefore delivered him into the hand of the mighty one of the heathen; he shall surely deal with him: I have driven him out for his wickedness.**

The word for height in the phrase "thyself in height" is the same word used to provide the ark's third dimension; and so we see the word used for the "height" of the ark to be associated with man's pride.

Let's incorporate these three associations from the ark's dimensions into our original statement of what God may be using the numbers to tell us. Based only on the numbers, we may say that it is God's complete purpose to apply the atonement, thereby bringing salvation and judgment. Incorporating the associations seen in the three words used to define the ark's dimensions, we may say that it is God's complete purpose, throughout all time and over all the earth, to apply the atonement (to bring salvation and judgment) - as a consequence of man's pride, which is wickedness.

Note that if we had converted the cubit to another unit of measurement (as done in some translations of the Bible), thereby changing the numbers God has used to define the ark, we would lose any possibility of understanding the spiritual lesson God has hidden in the scripture containing the ark's dimensions.

Let us now list all the numbers of spiritual significance previously discussed, with the corresponding meaning that each number is understood to have:

2- Those who have been commissioned to bring the Gospel

3- God's purpose

4- The farthest extent in time or in distance that God spiritually has in view

5- The atonement, which emphasizes both judgment and salvation

7- The perfect fulfillment of God's purpose

10- The completeness of whatever is in view

11- The first coming of Christ, 11,000 years after creation

12- The fullness of whatever is in view

13- The end of the world time period, beginning exactly 13,000 years after creation

17- Heaven

23- God's wrath or judgment

37- God's wrath or judgment
40- Testing
43- God's wrath or judgment

We have seen that these numbers have spiritual significance. We will find that by factoring certain numbers into these numbers of spiritual significance, we have an amazing tool for analyzing time intervals between major dates derived from the Biblical calendar.

The Calendar Patriarchs

When we read the first chapter of Genesis, we learn that it was on the sixth day that God created man. In Genesis 1:27, we read:

> So God created man in his *own* image, in the image of God created he him; male and female created he them.

Adam was created on the sixth literal day after God began His work of creation. The universe was less than a week old, and everything needed to sustain human life was in place on earth. That's when God created man. As we continue reading the Book of Genesis, we find that there is continuity of time. There are no gaps of time for which we cannot account. In Genesis 5:2-5, we read:

> Male and female created he them; and blessed them, and called their name Adam, in the day when they were created. And Adam lived an hundred and thirty years, and begat *a son* in his own likeness, after his image; and called his name Seth: And the days of Adam after he had begotten Seth were eight hundred years: and he begat sons and daughters: And all the days that Adam lived were nine hundred and thirty years: and he died.

From Adam, and continuing on through a long series of patriarchs - whom we may call calendar patriarchs - God builds a calendar. A great deal of this calendar is contained in Genesis chapters 5 and 11. In Genesis 11:16-17, we see a typical formula for presenting the information about one of these calendar patriarchs, in this case a man named Eber:

> **And Eber lived four and thirty years, and begat Peleg: And Eber lived after he begat Peleg four hundred and thirty years, and begat sons and daughters.**

As we read this scripture about Eber and Peleg, and compare it with the previously quoted scripture about Adam and Seth, we might easily conclude that Peleg was Eber's son. Seth was indeed Adam's son. Notice, however, that Adam named his son: he "called his name Seth." We do not find comparable words in the scripture about Eber and Peleg. In fact, we cannot conclude that Peleg was the son of Eber.

What we can conclude is that a child was born to Eber when he was 34 years old, and that this child was an ancestor of Peleg. How many generations there were between Eber and Peleg we don't know; but through the child that Eber bore, Peleg was begotten by Eber. Why, then, are we told about Peleg instead of another of Eber's descendants? For this reason: Peleg was born in the year Eber died. In this way, we have continuity of time.

How can we know that this method of interpreting the scripture about Eber and Peleg is correct? God has provided a way to confirm that it is. In Exodus 12:40-41, we read that the children of Israel were in Egypt for 430 years:

> **Now the sojourning of the children of Israel, who dwelt in Egypt, *was* four hundred and thirty years. And it came to pass at he end of the four hundred and thirty years, even the selfsame day it came to pass, that all the hosts of the LORD went out from the land of Egypt.**

The patriarchs associated with this time in Egypt are Levi (who went into Egypt as one of the original settlers), Kohath and Amram (who lived their entire lives there), and Aaron (who left Egypt in the Exodus). When we add up the years the Bible gives for these men, interpreting the numbers as we did for Eber and Peleg, we get a perfect tally of 430 years. In this way, we know we are on the right track in our understanding of the calendar patriarch method

Chapter 6 God's Calendar

God is using.

To resume our discussion of the calendar information contained in Genesis chapters 5 and 11, we find that when we go as far as the end of Nahor's period and add up the numbers, we get a total of 8,716 years. Within this number are the years for the lives of Noah and his son Shem. God presents their time information in a somewhat different manner because the flood occurred during Noah's day and only those who were on the ark survived. (The flood occurred 2,693 years earlier than the end of Nahor's period, in the year 6,023 according to the Biblical calendar.)

Following Nahor as a calendar patriarch is Terah - Abram's father; then come Abram (whom God renamed Abraham), Isaac and Jacob. The stories of these men continue well past chapter 11 of Genesis. For these men, God presents the time information in a different way than He does for Eber and Peleg. The time information for Abraham, Isaac and Jacob is woven into the stories of their lives.

The calendar continues with the birth of Abram to Terah. By comparing scriptures from Genesis 11 and 12 with each other, we can determine that Terah was 130 years old when Abram was born. This occurred in the year 8,846 of the Biblical calendar.

The next event to advance the calendar is the circumcision of Abraham, as he was now called. Abraham was 99 years old at that time, and so we come to the Biblical calendar year 8,945.

In Genesis 21, we read of the birth of Abraham's son Isaac. Abraham was 100 years old at the time. Therefore, Isaac was born in the year 8,946 of the Biblical calendar. Isaac's son Jacob was born when Isaac was 60 years old. This event, which we read about in Genesis 25, advances the Biblical calendar to the year 9,006.

Part of Jacob's story is his trek into Egypt; there his family grew into a great people, although they ended up enslaved by the Egyptians. Jacob was 130 years old when he entered Egypt. This event, which we read about in Genesis 47, occurred in the year 9,136 of the Biblical calendar.

In Exodus 12, we read that the children of Israel (God gave this name to Jacob) were in Egypt for 430 years. When we add this number to our running total, we find that 9,566 years have passed from the time of Creation until the Exodus.

About Carbon 14 Testing

At this point, a major issue needs to be addressed. How can we trust the Bible when it tells us that only 9,566 years elapsed from the Creation until the Exodus? Don't we know from science, beyond the shadow of a doubt, that the earth is billions of years old, and that life has existed on the earth for many millions of years? This is what is commonly taught in our schools. It is a belief that permeates our culture. If you watch a nature program on television, there is a strong likelihood that this view will be presented as fact - not as theory. You need to know, however, that it is not a fact - it is a theory, and it's all wrong.

Whenever a scientific dating method is used, one or more *assumptions* must be made. If the assumption is wrong, the result is wrong. For example, carbon 14 testing is a major tool used for dating objects that are up to 50,000 years old. At least, it is considered accurate for testing something as old as that. It has been used successfully - that is, the result of carbon 14 testing has been confirmed by historical records. However, the earliest known writing only goes back to about 3500 B.C., so it would never be possible to verify the result of carbon 14 testing from the historical record for anything earlier.

We should take a brief look at carbon 14 testing at this point. Carbon 14 is an isotope - that is, an unstable form - of the common carbon atom. It results from radiation hitting the earth's atmosphere. Even though carbon 14 is unstable, it tends to "hang around" for quite a while. Its rate of decay is such that it takes about 5,730 years for one half of a given amount of it to change into nitrogen (the other half remains and continues to decay). This amount of time is called the half life of carbon 14.

Carbon 14 bonds with oxygen to become a radioactive form

of carbon dioxide. This radioactive form of carbon dioxide is all over the place, but only in minuscule concentrations. You may have heard the expression, "you are what you eat." Well, the food we eat and the water we drink contain this carbon 14. We keep ingesting it as long as we live. Scientists can measure it today in a living thing, so they have an idea of how much there should be, for example, in the bones of someone who recently died or in a piece of wood that was recently chopped off a tree. When the person or the tree dies, the ingestion of carbon 14 stops. Then, whatever carbon 14 is in the person or the tree begins decaying with a half life of about 5,730 years. So if a scientist tests a bone fragment of someone who died many years ago and finds it contains half as much as it should - compared to what a living organism contains - the scientist concludes that the person died about 5,730 years ago. There's a nice mathematical equation that allows the scientist to plug in the numbers and calculate the sample's age. A scientist may adjust the result of the equation using graphs or other equations, but that is the basic idea behind carbon 14 testing.

The amount of carbon 14 in the environment has been compared to the amount of water in a leaking barrel. The barrel is being filled by a hose (as carbon 14 is produced) and it's leaking because it has holes (as carbon 14 decays). If the barrel is full, then the amount of carbon 14 in the environment is steady. Suppose the "barrel became full" thousands of years ago and continued that way up to the present. This means that someone who was alive thousands of years ago was ingesting the same amount of carbon 14 from his environment as we do today. Suppose, however, that this is not the case. (This consideration also applies to tests using other types of radioactive elements - such as uranium - having a much longer half life.) In other words, if the "barrel" is still "filling up," then we don't have a steady reference level from the past with which we can compare carbon 14 concentrations in today's living organisms.

There are other considerations as well. It is known that nuclear testing conducted years ago affects the results of carbon 14 testing. As a result of nuclear testing, we may find higher concentrations of carbon 14 in the northern hemisphere than in the southern hemisphere. The burning of fossil fuels is another problem for carbon 14 testing. Not all fossil fuels affect carbon 14 test results in the same way. Some fossil fuels may be especially rich in carbon 14; in other fossil fuels, the amount of carbon 14 may be especially small. Depending on the fuels that have been consumed in any given area and when they have been consumed, the result of carbon 14 testing on a sample from that area may be affected so that carbon 14 testing will either underestimate or overestimate its age.

There are yet other problems that scientists either aren't aware of, or won't acknowledge. Let's suppose the Bible is correct in telling us that there was a global flood in the days of Noah, and that this flood occurred almost 7,000 years ago. (We will see that this is when it did occur, according to the Bible's calendar.) We should expect that such a cataclysmic event would drastically affect the production of carbon 14. Read what the Bible tells us about this flood in Genesis 7:11-12:

> **In the six hundredth year of Noah's life, in the second month, the seventeenth day of the month, the same day were all the fountains of the great deep broken up, and the windows of heaven were opened. And the rain was upon the earth forty days and forty nights.**

The Bible tells us that the high hills and the mountains were actually covered in this flood. The average temperature around the planet should have been considerably lower for many years afterwards as a result, precipitating an ice age. All of this could have drastically reduced the amount of carbon 14 in the environment.

Let's imagine we have a bone of someone who lived around that time, about 7,000 years ago, and we bring it in for carbon 14 testing. The scientist will be looking for a certain amount of carbon 14 in order to tell us that the bone is 7,000 years old. Because of what happened to the earth at that time, however, his test may reveal that the bone contains much less carbon 14 than that. The equation will therefore tell him that the sample is much more than

Chapter 6 God's Calendar

7,000 years old.

The Bible also gives us a hint of another incredible phenomenon in earth's history. In Genesis 10:25, we read:

And unto Eber were born two sons: the name of one *was* **Peleg; for in his days was the earth divided; and his brother's name** *was* **Joktan.**

Science tells us that there was at one time a supercontinent called Pangaea (Greek for "all land"), and that millions of years ago this land mass broke up to eventually form the continents of today. Could the words "for in his days was the earth divided" in this scripture be referring to that event? When we work through the Bible's calendar, we find that Peleg lived about 5,000 years ago. If that's when the continents broke up, there may have been an enormous number of earthquakes and tremendous volcanic activity around that time. By the standards of modern science, the continental breakup would have occurred with breathtaking speed. It is reasonable to assume that the resulting volcanic eruptions and earthquakes affected the amount of carbon 14 in the environment, possibly increasing it tremendously in some areas.

Let us now imagine we have a bone of someone who lived around the time this continental drift began. We bring this bone in for carbon 14 testing. The scientist will be looking for a certain amount of carbon 14 in order to tell us that the bone is 5,000 years old. Because of what happened on earth at that time, however, his test may reveal that the bone contains much more carbon 14 than that. His equation will therefore tell him that the sample is much less than 5,000 years old.

There's another problem with carbon 14 testing, and it's a major one. There is an underlying assumption that a sample (of bone or wood, for example) is not affected by the carbon 14 in its environment after the organism has died. Carbon 14 testing assumes that the sample is a "vault" - that there is no exchange of carbon 14 between the sample and its environment after a certain time. In some situations, however, that sample may be more like a sponge than a vault. The sample's situation (e.g., buried in the earth or stored in a clay jar that has been buried) and the physical history of that location could drastically affect its carbon 14 content.

We have seen that carbon 14 testing yields results that cannot be blindly accepted. Mankind's incredible accomplishments in space travel, computers, communications and other areas do not mean that we can know the past. The Bible, on the other hand, has proven itself to be the word of God. When we understand what it tells us about the past, we know we can depend on it as the truth.

Continuing With The Biblical Calendar

Let us now return to the Biblical Calendar. We have seen that 9,566 years elapsed from the time of Creation until the Exodus. The Bible tells us that the children of Israel wandered in the wilderness for 40 years after they left Egypt, as we read in Joshua 5:6:

> **For the children of Israel walked forty years in the wilderness, till all the people *that were* men of war, which came out of Egypt, were consumed, because they obeyed not the voice of the LORD: unto whom the LORD sware that he would not shew them the land, which the LORD sware unto their fathers that he would give us, a land that floweth with milk and honey.**

Moses led them in the wilderness until his death, shortly before they crossed the Jordan River into the land of Canaan; but it was Joshua who led them across, as God parted the waters so they could pass over on dry ground.

The first year that the children of Israel were in the land of Canaan was to be kept as a Sabbath year. The date of this first Sabbath year, which occurred forty years after the Exodus, is the year 9,606 according to the Biblical calendar. The first Jubilee year occurred fifty years after that, in the year 9,656 according to the Biblical calendar. Later on, we will discuss the significance of this. The idea of the Jubilee assists us in understanding the Biblical calendar and the significance of the annual feasts in God's

Chapter 6 God's Calendar

salvation plan.

The Book of Joshua gives us an account of Israel's conquest of the land of Canaan under Joshua after they crossed the Jordan. By the end of Joshua's life, the children of Israel had divided up the land. He cautioned the children of Israel to be faithful to the Lord and then let them go, as we read in Joshua 24:28-29:

> **So Joshua let the people depart, every man unto his inheritance. And it came to pass after these things, that Joshua the son of Nun, the servant of the LORD, died, *being* an hundred and ten years old.**

After the death of Joshua, God raised up a series of men (and a woman) to deliver the children of Israel from their enemies. We read about these leaders in the Book of Judges. The children of Israel repeatedly turned away from the Lord, and so He allowed their enemies to gain the upper hand over them. Eventually, after they cried out to God in their suffering, He delivered them by using one of these judges. This cycle is repeated many times in the Book of Judges.

The last of the Judges was a man named Samuel. It was he whom God directed to anoint a king for Israel. Israel's first king was Saul. He was followed by David as king, and then David's son Solomon succeeded him as king. It was during Solomon's reign that the temple, called the house of the LORD in 1 Kings 6:1, was built:

> **And it came to pass in the four hundred and eightieth year after the children of Israel were come out of the land of Egypt, in the fourth year of Solomon's reign over Israel, in the month Zif, which *is* the second month, that he began to build the house of the LORD.**

This period of 480 years includes the years that Moses and Joshua led the children of Israel, the period of the judges, the reigns of Saul and David, and four years of Solomon's reign. Adding 480 to the 9,566 years that passed from the time of Creation until the Exodus, we get 10,046 years. Notice, however, that this brings us to the fourth year of Solomon's reign. The Bible tells us that

Solomon reigned for 40 years; so if we add 36 to our running total, we find that 10,082 years have passed at the end of Solomon's reign.

Solomon was the third and last king to rule all of Israel - referred to as the united monarchy - for his entire reign. Solomon's son Rehoboam succeeded him as king, but ten of the tribes were taken away from him by God and given to a man named Jeroboam. Rehoboam's kingdom came to be known as Judah; it consisted of only two tribes: Judah and Benjamin. Jeroboam's kingdom was known as Israel. This splitting of the kingdom occurred in the year 10,082 of the Biblical calendar.

Let us continue to follow the time path by briefly considering the history of the kingdom of Judah. After Rehoboam, the kingdom of Judah had 19 more kings. In the Bible, the books called 1 Kings, 2 Kings, 1 Chronicles and 2 Chronicles contain much detailed information about the kings of both Judah and Israel. These books provide the time information needed to determine how long each king reigned.

Judah's twentieth and last king was Zedekiah. His last year as king was the year the Babylonians, under king Nebuchadnezzar, captured Jerusalem and destroyed the temple. If we add up the number of years that Judah's kings reigned - from the first year of Rehoboam to the last year of Zedekiah - we get a total of 344 years. When we add this number to the Biblical calendar date we calculated for the splitting of the united kingdom (10,082) in Rehoboam's first year, we arrive at the year 10,426 according to the Biblical calendar. The Biblical calendar date for the fall of Jerusalem to Babylon is therefore 10,426.

The Bible tells us that Zedekiah's reign lasted for 11 years. In fact, Zedekiah was made king of Judah by Nebuchadnezzar, as we read in 2 Kings 24:17:

And the king of Babylon made Mattaniah his father's brother king in his stead, and changed his name to Zedekiah.

The king of Babylon carried away Judah's king Jehoiachin into

Chapter 6 God's Calendar

captivity and made Jehoiachin's uncle the new king, naming him Zedekiah. It was because of Zedekiah's rebellion against Babylon that Nebuchadnezzar sent his forces again and eventually destroyed Jerusalem. Later, we will see that Zedekiah's first year (10,416 in the Biblical calendar) is a helpful milestone.

Now let us take a brief look at the kingdom of Israel in the years following the split in the united monarchy. Israel had twenty kings, from Jeroboam to Hoshea inclusive. They reigned a total of 222 years. When we add this number to the Biblical calendar date for the splitting of the united monarchy, we arrive at the year 10,304 in the Biblical calendar. That was the year the Assyrians conquered Samaria, Israel's capital, thereby ending the kingdom of Israel.

One of Israel's twenty kings is of particular interest: that is King Ahab. He was Israel's eighth king after the split in the united monarchy, and he began his reign 57 years after Jeroboam's first year. Adding 57 to the year that Jeroboam began his reign (10,082) brings us to the Biblical calendar year 10,139 as Ahab's first year. The Bible tells us that Ahab reigned for 22 years; his last year is therefore 10,160 according to the Biblical calendar. This date is a second milestone that allows us to synchronize the Biblical calendar with the modern calendar.

Synchronizing the Calendars

If you were to do a little checking in history books dealing with ancient Israel, you will most likely find the date 597 B.C. as Zedekiah's first year, and 853 B.C. as Ahab's last year. These dates, although not universally accepted, are commonly accepted based on sources outside the Bible. Notice the Biblical calendar's dates for Zedekiah's first year and Ahab's last year, and compare them with the corresponding dates according to the modern calendar:

Zedekiah (first year)   10,416 (Biblical calendar)   597 B.C.

Ahab (last year)   10,160 (Biblical calendar)   853 B.C.

According to the Biblical calendar, Ahab's last year was 256 years earlier than Zedekiah's first year. This is the same difference between the modern calendar's dates! This means that we can synchronize the Biblical calendar with our modern calendar.

Up to this point, we have been using the Bible's own calendar to establish dates for Biblical events. Now we know that a Biblical calendar date of 10,416 corresponds to the year 597 B.C. so we can state Biblical dates in terms of the modern calendar by simply adding or subtracting time intervals. For instance, we arrived at the Biblical year 9,656 for the first Jubilee year. Using our benchmark Biblical year of 10,416, we find a difference of 760 years:

10,416 - 9,656 = 760

Therefore, the first Jubilee year occurred 760 years earlier in history than our benchmark year. This translates to:

760 + 597 = 1,357   (or: - 760 - 597 = -1,357)

So we come to the date 1357 B.C. as the first Jubilee Year according to our calendar. In this manner we arrive at the following dates for the Biblical events we discussed:

Creation	11,013 B.C.
The Flood	4990 B.C.
End of Nahor's period	2297 B.C.
Birth of Abram	2167 B.C.
Circumcision of Abraham	2068 B.C.
Birth of Isaac	2067 B.C.
Birth of Jacob	2007 B.C.
Jacob enters Egypt	1877 B.C.
Exodus from Egypt	1447 B.C.
First Sabbath Year	1407 B.C.

Chapter 6 God's Calendar

(Israel enters promised land)
First Jubilee Year   1357 B.C.
Temple construction begun   967 B.C.
Splitting of united monarchy   931 B.C.
Fall of Samaria   709 B.C.
Fall of Jerusalem   587 B.C.

Now that we have synchronized the Biblical calendar with our modern calendar, we may use the modern calendar when referring to Biblical events.

It should be mentioned that a commonly accepted date for the Fall of Samaria to Assyria - and the end of the nation of Israel - is 722 B.C., but the Bible does not support that date. The 722 B.C. date is based solely on archaeological evidence. The archaeological record, however, is incomplete. Furthermore, it has undoubtedly been tampered with in many instances. We need to realize that the destruction of monuments and obliteration of inscriptions on monuments was not limited to the effects of time and vandalism: a vain king could easily have ordered an inscription to be carefully chiseled away, in order to diminish his predecessor's reputation and enhance his own.

The Bible tells us that Israel's last king was named Hoshea. When we add up the years for all the kings who followed king Jeroboam up to the time of king Hoshea, we find that Hoshea's reign began in 718 B.C., which was 213 years after Jeroboam began to reign. The Bible also tells us that the final siege against Israel began in the seventh year of Hoshea's reign, when Hezekiah was king of Judah. In 2 Kings 18:9-10, we read:

> **And it came to pass in the fourth year of king Hezekiah, which *was* the seventh year of Hoshea son of Elah king of Israel, *that* Shalmaneser king of Assyria came up against Samaria, and besieged it. And at the end of three years they took it: *even* in the sixth year of**

Hezekiah, that *is* the ninth year of Hoshea king of Israel, Samaria was taken.

Based on inscriptions found in ancient ruins, archaeologists have concluded that Shalmaneser was king from 727 B.C. to 722 B.C. - a reign of only five years. According to the Bible, however, Shalmaneser was still king as late as 711 B.C., which was the seventh year of Hoshea. The Assyrian king Sargon, who claimed the victory over Israel, apparently tampered with the historical record in order to glorify himself at the expense of Shalmaneser. Whenever there is a discrepancy between a Biblical account and archaeological evidence, we know that the Bible is where we may find the truth.

Following the end of the kingdom of Israel, the kingdom of Judah managed to survive for another 122 years before it too was conquered. One might expect that, with the end of Judah, there would be no way to continue following the Biblical timeline of history. It is true that, from this time on, the Bible no longer provides a continuous calendar as it does in Genesis chapters 5 and 11. However, by carefully comparing the scriptures that do provide time information with the historical record, we are able to date key events following the fall of Judah.

After the Fall of Judah

In the Book of Jeremiah, there is an astonishing prophecy. Jeremiah had been sent to warn the kingdom of Judah to repent. They never did, and so God decided to bring His judgment against them. The Babylonians, also called the Chaldeans, were God's instrument to do this. In Jeremiah 25:11-12, we read this prophecy - a dreadful announcement which Jeremiah delivered to Judah:

> **And this whole nation shall be a desolation, *and* an astonishment; and these nations shall serve the king of Babylon seventy years. And it shall come to pass, when seventy years are accomplished, *that* I will punish the king of Babylon, and that nation, saith the LORD, for their iniquity, and the land of the Chaldeans, and will**

make it perpetual desolations.

Throughout the centuries that scholars have been studying the Bible, this prophecy has undoubtedly been much analyzed and its meaning much debated. Undoubtedly, there is spiritual meaning to these verses: so much of the Bible has spiritual meaning - that is, a meaning outside of the meaning we would get if we read the words as though they were coming from an ordinary book. We will find, however, that when we apply this prophecy in a certain way, we can actually use the seventy year figure as a time bridge that allows us to continue with the Biblical timeline; but before we get to that step, we first need to backtrack.

The fall of Jerusalem was not an instantaneous affair. The kingdom of Judah had been weakened over the course of many years. Judah's last good king was named Josiah, and as early as the end of his reign - 22 years before the fall of Jerusalem - the end of Judah was in sight. When King Josiah was killed in battle against Egypt, his son Jehoahaz was anointed by the people and became the new king. Only three months into his reign, however, the Egyptian pharaoh removed him and made another of Josiah's sons the king. We read about this in 2 Kings 23:33-34:

> **And Pharaohnechoh put him in bands at Riblah in the land of Hamath, that he might not reign in Jerusalem; and put the land to a tribute of an hundred talents of silver, and a talent of gold. And Pharaohnechoh made Eliakim the son of Josiah king in the room of Josiah his father, and turned his name to Jehoiakim, and took Jehoahaz away: and he came to Egypt, and died there.**

It was the year 609 B.C. that the Egyptian pharaoh removed Judah's king Jehoahaz and replaced him with someone of his own choosing. The kingdom of Judah was never really independent after that.

When we first read Jeremiah's prophecy about the 70 years of servitude to Babylon, it may not be clear that this period of time began in the year that the Egyptian pharaoh removed Jehoahaz. God, however, did not make the Bible easy to understand. God tells us that at the end of the 70 years, He would punish the king of Babylon. Sure enough, when we add 70 years to the year 609 B.C., we come to the year 539 B.C. - the year that Cyrus of Persia conquered Babylon according to the history books.

The prophecy of the 70 years is like a bridge connecting the time of Judah's kings to a time of great Media-Persian kings, beginning with Cyrus. As we read the books of the Bible that deal with these kings, we find that the names the Bible uses for them presents a difficulty. Some of these names were most likely titles. Additionally, we need to realize that a king may have been known by more than one name. For example, Cyrus probably found it helpful to be known as King Cyrus of Persia when dealing with his Persian subjects, and as King Darius the Mede when dealing with his original countrymen. Despite these complications, by carefully comparing the historical record with the Biblical account, the identities of these kings has been established.

The Bible calls 539 B.C. Cyrus' first year. Although history tells us that Cyrus actually became king in 559 B.C. rather than 539 B.C., there is no disagreement with the Bible here. From the perspective of the Jews who were under Babylonian rule, the year that Cyrus took over the Babylonian kingdom was his first year. It was also his first year as king of a greatly expanded empire, and so we read in 2 Chronicles 36:22-23:

> **Now in the first year of Cyrus king of Persia, that the word of the LORD *spoken* by the mouth of Jeremiah might be accomplished, the LORD stirred up the spirit of Cyrus king of Persia, that he made a proclamation throughout all his kingdom, and *put it* also in writing, saying, Thus saith Cyrus king of Persia, All the kingdoms of the earth hath the LORD God of heaven given me; and he hath charged me to build him an house in Jerusalem, which *is* in Judah. Who *is there* among you of all his people? The LORD his God *be* with him, and let him go up.**

God raised up Cyrus for His own purpose. In fact, there is even a prophecy concerning Cyrus. In the prophecy, written years before

Chapter 6 God's Calendar

Cyrus was even born, Cyrus is called God's anointed and His shepherd. This prophecy is found in Isaiah 44:28-45:1:

> That saith of Cyrus, *He is* my shepherd, and shall perform all my pleasure: even saying to Jerusalem, Thou shalt be built; and to the temple, Thy foundation shall be laid. Thus saith the LORD to his anointed, to Cyrus, whose right hand I have holden, to subdue nations before him; and I will loose the loins of kings, to open before him the two leaved gates; and the gates shall not be shut;

In these verses, we see Cyrus as a picture of the Lord Jesus. Now that Cyrus was king of the lands where the Jews - who had been relocated by the Babylonians - were living, he was able to permit them to return to their homeland and begin building a new temple. It took years for this temple to be finished, but Cyrus' proclamation was the start.

Cyrus ruled until 529 B.C. - the year he was killed in battle according to the archaeological record. His son Cambyses, known as Cambyses II, succeeded him. The Bible tells us that during Cambyses' reign the enemies of the Jews managed to stop work on the temple. In the Bible, this king is called Artaxerxes and Ahasuerus. Cambyses died in 522 B.C.

The archaeological record tells us that Cambyses was succeeded by Darius I (who ruled from 522 B.C. to 485 B.C. and is known to history as Darius the Great). There was a usurper who took the name Smerdis and ruled briefly in 522 B.C., but Darius killed him. During the reign of Darius, the temple was finally completed. We read about this in Ezra 6:15:

> **And this house was finished on the third day of the month Adar, which was in the sixth year of the reign of Darius the king.**

It's important to note something about the given dates during which these kings of Media-Persia reigned: the first year is the year of accession, so that the next year is actually the first full year of their reign. This would make 515 B.C. the "sixth year of the reign of Darius the king," the year the temple was finished.

Continuing with the archaeological record, the next king to rule was Xerxes I. His reign is from 485 B.C. to 465 B.C.

Xerxes I was succeeded by his son Artaxerxes I, who ruled from 465 B.C. to 424 B.C. and is known to history as Artaxerxes II Longimanus. During the reign of this king, Ezra the scribe returned to Jerusalem, as we read in Ezra 7:6-8:

> **This Ezra went up from Babylon; and he *was* a ready scribe in the law of Moses, which the LORD God of Israel had given: and the king granted him all his request, according to the hand of the LORD his God upon him. And there went up *some* of the children of Israel, and of the priests, and the Levites, and the singers, and the porters, and the Nethinims, unto Jerusalem, in the seventh year of Artaxerxes the king. And he came to Jerusalem in the fifth month, which was in the seventh year of the king.**

Later on, we will see that the year of Ezra's return to Jerusalem - 458 B.C., being the seventh year of this king Artaxerxes - is extremely significant. In Ezra 7:10, we see why Ezra had made his request to the king:

> **For Ezra had prepared his heart to seek the law of the LORD, and to do *it*, and to teach in Israel statutes and judgments.**

In this chapter of the Bible, we also learn the contents of a letter the king gave to Ezra. This amazing letter authorized Ezra to bring gold and silver that had been provided by the king and his counselors, to collect freewill offerings of the people, and to receive supplies from the king's treasurers. Failure to cooperate with Ezra could result in severe penalties, even death. The funds were to be used to buy meat offerings and drink offerings prescribed by God's laws for the temple sacrifices, and to do whatever else was needed. In other words, Ezra went back to Jerusalem in 458 B.C. to reestablish God's law.

The two kings who followed King Artaxerxes I were Xerxes II (who reigned for only two months in 424 B.C. before he was murdered) and Darius II (who reigned from 424 B.C. to 404 B.C.). There is nothing in the Bible to identify these two kings.

The next king to reign was Artaxerxes II. His reign is dated from 404 B.C. to 358 B.C., and he has been identified as the King Ahasueurus of the Book of Esther. The Book of Esther is not a fable, but a record of actual events. Our interest is to determine the time setting of this book.

Shortly after the story begins, we find the wicked Haman plotting to destroy the Jews. In Esther 3:7, we read:

> **In the first month, that *is*, the month Nisan, in the twelfth year of king Ahasuerus, they cast Pur, that *is*, the lot, before Haman from day to day, and from month to month, *to* the twelfth *month*, that *is*, the month Adar.**

Here we learn that the story is set in the twelfth year of the king's reign. Haman's plan - to use the king's power to murder the Jews - was set to be executed in the twelfth month of that year, as we learn by reading the story and as we see in Esther 9:1:

> **Now in the twelfth month, that *is*, the month Adar, on the thirteenth day of the same, when the king's commandment and his decree drew near to be put in execution, in the day that the enemies of the Jews hoped to have power over them, (though it was turned to the contrary, that the Jews had rule over them that hated them;)**

The culmination of this book occurs in the twelfth month of the twelfth year of the king, which we know to be 391 B.C. by matching the scriptures containing time information against the historical record telling us that 404 B.C. is the year this king began to reign. Later on, we will consider this 391 B.C. date as it relates to dates for other Biblical events.

The Book of Esther is a historical parable pointing to the end of the world, when Satan (identified in the story with Haman, the king's highly placed prince) meets his end. The Book of Esther is also significant because there is no evidence of anything else in the Old Testament occurring after 391 B.C., making this the last word from God until He gave us the New Testament.

The Timeline Continues to the New Testament - The Birth of the Lord

The next major event we will consider is the birth of the Lord Jesus. Our modern calendar uses the Anno Domini method of designating the current year. These words, abbreviated as A.D., mean "year of the Lord" in Latin and refer to the year in which the Lord Jesus was born. We might, therefore, reasonably expect that we should know the year of the Lord's birth by virtue of this method of keeping our calendar. After all, this method has been commonly accepted - at least in Europe - for over a thousand years. It was developed almost 1500 years ago by a monk known to history as Dionysius Exiguus. When he developed it, he most likely had in mind either 1 B.C. or 1 A.D. as the date of our Lord's birth. We shall see that he was off by a few years.

The Bible does not appear to allow us to determine a date for the Lord's birth the way it allows us to determine dates for major events in the Old Testament, as we previously did. However, we can actually arrive at a date when we consider the Old Testament's special days and years that were to be observed as part of the law given to ancient Israel, and realize that there is a connection between them and New Testament events. Once we arrive at the date, we will see how the Bible confirms it in a variety of ways.

The Bible clearly tells us that two major New Testament events occurred on annual feast days: first, that the Lord Jesus was crucified on Passover; and second, that the Holy Spirit was poured out on Pentecost to begin the period when God used local congregations to bring the Gospel. The birth of our Lord certainly ranks as a major New Testament event. If we remember that the Lord Jesus used language suggesting the year of Jubilee when

Chapter 6 God's Calendar

speaking of His ministry, we are on the right track to determine His birth date.

Earlier, we saw that the first year of Jubilee occurred fifty years after the children of Israel entered the promised land. To review briefly, the Exodus was in 1447 B.C., Israel spent forty years in the wilderness before they entered the promised land in 1407 B.C., and in 1357 B.C. - fifty years later - they observed their first Jubilee year. Thereafter, every fifty years was a Jubilee year. Therefore, the year 7 B.C. was a Jubilee year. Based on a statement the Lord Jesus made, associating Himself with the Jubilee, we focus on 7 B.C. as the year of His birth. The Bible tells us that Herod was king when the Lord Jesus was born. In Matthew 2:1-2, we read about this in connection with visitors referred to as wise men:

> **Now when Jesus was born in Bethlehem of Judaea in the days of Herod the king, behold, there came wise men from the east to Jerusalem, Saying, Where is he that is born King of the Jews? for we have seen his star in the east, and are come to worship him.**

When we check an encyclopedia, we find that there were several kings named Herod. The one who was alive in the year 7 B.C. is known as Herod the Great. According to history, he died in April of 4 B.C. - a date that is considered reliable and is consistent with our 7 B.C. date for the birth of the Lord.

Herod had asked the wise men to inform him of the child's location; but the wise men were warned by God not to do so, and so they returned by another route to their own country after they had been guided to the Lord Jesus by a miraculous star. Herod was furious that the wise men did not report back to him. Apparently he was driven by an insane compulsion to perpetuate his name, to guarantee that the throne would pass to his offspring, and to protect it against the child who was born King of the Jews - Herod's own title. Herod issued the order to kill all male children in the region, hoping to kill the one child who might some day threaten his house. We read about this in Matthew 2:16:

> **Then Herod, when he saw that he was mocked of the wise men, was exceeding wroth, and sent forth, and slew all the children that were in Bethlehem, and in all the coasts thereof, from two years old and under, according to the time which he had diligently inquired of the wise men.**

The 4 B.C. date for Herod's death does not confirm our date for the birth of the Lord, but it's consistent with it. This is so because the Bible tells us that the children Herod ordered to be killed were two years old and under. When he gave this order, he must have believed that no more than two years had passed since the time the child was born. Herod was still alive in April of 4 B.C., more than two years after the Lord's birth, to give this order; and so we have consistency with the secular record and the scripture telling us about this order.

The Bible tells us that the wise men were guided by a miraculous star to the exact house where the child Jesus was. They brought Him gifts of gold, frankincense and myrrh. When the wise men left to return to their own county, God sent an angel to Joseph, warning him to flee to Egypt. We read about this in Matthew 2:13-15:

> **And when they were departed, behold, the angel of the Lord appeareth to Joseph in a dream, saying, Arise, and take the young child and his mother, and flee into Egypt, and be thou there until I bring thee word: for Herod will seek the young child to destroy him. When he arose, he took the young child and his mother by night, and departed into Egypt: And was there until the death of Herod: that it might be fulfilled which was spoken of the Lord by the prophet, saying, Out of Egypt have I called my son.**

The Bible also gives us the date when the Lord's ministry began. We know from history that 29 A.D. was the fifteenth year of Tiberius. In Luke 3:1-2, we read:

> **Now in the fifteenth year of the reign of Tiberius**

Chapter 6 God's Calendar

> Caesar, Pontius Pilate being governor of Judaea, and Herod being tetrarch of Galilee, and his brother Philip tetrarch of Ituraea and of the region of Trachonitis, and Lysanias the tetrarch of Abilene, Annas and Caiaphas being the high priests, the word of God came unto John the son of Zacharias in the wilderness.

Later in the third chapter of Luke, and based on what we learn in the other Gospel accounts, we find that this John - known as John the Baptist, who is not to be confused with the apostle John - baptized the Lord Jesus. Shortly after this, the Lord began His ministry. All of this appears to have happened in the fifteenth year of the reign of the Roman emperor Tiberius. Tiberius ascended the throne in 14 A.D., and 15 A.D. was his first full year of reign. Therefore, counting from 15 A.D. as the first year, we get to 29 A.D. as the fifteenth year and the time setting for these events. In Luke 3:23, we're apparently given the Lord's approximate age at that time:

> And Jesus himself began to be about thirty years of age, being (as was supposed) the son of Joseph, which was *the son* of Heli,

A careful look at the Greek, however, indicates that this scripture, is telling us something altogether different than the Lord's approximate age. It appears to be a reference to the year the Lord began to live in Israel - about thirty years earlier - after Joseph brought his family back from Egypt. We read about their return in Matthew 2:19-21:

> But when Herod was dead, behold, an angel of the Lord appeareth in a dream to Joseph in Egypt, Saying, Arise, and take the young child and his mother, and go into the land of Israel: for they are dead which sought the young child's life. And he arose, and took the young child and his mother, and came into the land of Israel.

We know that the Lord was actually 35 years old in the year 29 A.D., so He apparently began to live in Israel when He was about five years old.

Just as insight into the significance of the Jubilee year can lead us to the year of the Lord's birth, the annual feast days can lead us to the month and day. Earlier, we learned that a blast on the ram's horn on the Day of Atonement marked a Jubilee year. Also, it was only on this day that the high priest could enter into the most holy place in the temple, near the visible presence of the Lord. This was the Old Testament feast day which - according to a great deal of Biblical evidence - had its fulfillment in the birth of the Lord Jesus. The use of our English word "atonement" certainly seems appropriate here; for on the day He was born, the Lord Jesus was "at onement" or at one with us. He became the Son of Man just as He was already the Son of God.

The day on which the Day of Atonement fell in the year 7 B.C. has been calculated to be October 2. Although we haven't proved that this was the day on which the Lord was born, we will see later on how tightly this date fits into the overall framework of dates that we discover from the Bible.

There is actually Biblical support for a day near October 2 as being the day of the Lord's birth. Let's take a look at some of the evidence. We know that God used John the Baptist - Elisabeth's child - as a messenger for the Lord Jesus. The Bible gives us clues about the timing of John's birth, and we can use this information to establish the approximate time of the Lord's birth.

We know that an angel spoke to Mary, giving her the news of her impending conception. In Luke 1:35-36, we read:

> **And the angel answered and said unto her, The Holy Ghost shall come upon thee, and the power of the Highest shall overshadow thee: therefore also that holy thing which shall be born of thee shall be called the Son of God. And, behold, thy cousin Elisabeth, she hath also conceived a son in her old age: and this is the sixth month with her, who was called barren.**

As an estimate, let's say the angel made this announcement about half way through the sixth month of Elisabeth's pregnancy. Assuming that Mary was with child by the Holy Ghost almost immediately after this, and assuming that Elisabeth's and Mary's pregnancies were each of the normal nine month duration, we can conclude that Mary's child would be born about five and a half months after Elisabeth's.

The Bible gives us another clue - this one involving John the Baptist's father, who was a priest named Zacharias. In the first chapter of Luke, we learn that his "course" - that is, the period of time during which he performed his priestly duties - was "Abia." This turns out to be the last half of the fourth month of the Jewish calendar. The Bible also tells us that after he completed his course he returned to his house. After that, his wife conceived. Let's assume Zacharias arrived home early in the fifth month of the Jewish calendar. It has been calculated that, in 8 B.C., the last day of the fourth month of the Jewish calendar was early in our month of July, on July 5. Elisabeth may have conceived about the middle of that month, being about July 15. Therefore, John the Baptist would have been born the following year - about April 15 - and the Lord Jesus would have been born about five and a half months after that (about October 1, only one day earlier than our calculated date of October 2).

The Lord's Ministry Begins

The Bible tells us about an early event in the life of the Lord Jesus, before He began His ministry. He was twelve years old, and his parents took Him to Jerusalem for the Feast of the Passover. When His parents left to return home, they supposed He was in their group, but He had stayed behind. They came back to Jerusalem to look for Him, and were amazed to find Him in the temple - listening and asking questions of the learned men. Other than this event and the miraculous events involving His birth, the Bible doesn't tell us much about the early life of the Lord. Most of the four Gospels are concentrated on His ministry, about the last three and a half years of His life.

When exactly did the Lord's ministry begin? Previously, we saw that it began in the year 29 A.D., during the fifteenth year of

Tiberius. Amazingly, we can also discover the month and day the Lord's ministry began.

As we read the Gospel accounts, we learn that at the start of His ministry the Lord was tempted by Satan, that He gathered His disciples, and performed His first miracle in public at a marriage feast. Even before these things, however, something else occurred. In Matthew 3:13, we read:

Then cometh Jesus from Galilee to Jordan unto John, to be baptized of him.

People from all around Judaea came to John the Baptist to be baptized by him, and so did the Lord Jesus. In Matthew 3:16-17, we read what happened after the Lord was baptized:

And Jesus, when he was baptized, went up straightway out of the water: and, lo, the heavens were opened unto him, and he saw the Spirit of God descending like a dove, and lighting upon him: And lo a voice from heaven, saying, This is my beloved Son, in whom I am well pleased.

After seeing this, John the Baptist knew who the Lord was. In John 1:32, we read what he said:

And John bare record, saying, I saw the Spirit descending from heaven like a dove, and it abode upon him.

Around the time the Lord was baptized, John the Baptist announced Him as the Lamb of God. Earlier, we saw that there is an association between the Lord Jesus and the Jubilee year. We also saw that the Day of Atonement is related to the Jubilee. The feast known as the Feast of Trumpets is also related to the Jubilee because it was on that day every year that the ram's horn was sounded, as discussed earlier. We can see how beautifully the announcement by John the Baptist ties in with the sounding of the ram's horn. It is almost certain that the Lord was announced as the Lamb of God on the feast known as the Feast of Trumpets. Actually, John the Baptist made this announcement more than once; we read one instance of it in John 1:29:

> **The next day John seeth Jesus coming unto him, and saith, Behold the Lamb of God, which taketh away the sin of the world.**

This feast day, the first day of the seventh new moon of the year 29 A.D., occurred on September 26 according to calculation. What evidence might the Bible give us to support this date? We will see that the Bible indicates it by way of its relation to another major New Testament event - Pentecost. We will return to our discussion of the date of John the Baptist's announcement of the Lord after we briefly look at the final events in the Biblical timeline.

The Crucifixion, the Resurrection, and Pentecost

For about three and a half years, the Lord preached to multitudes of people, did many miracles, and taught His disciples. As we read the Gospels, we find that the Lord told His disciples what was coming. He told them He would die and rise from the dead, but they just couldn't understand. When the Lord was taken by the mob, humiliated, beaten, and finally crucified, His disciples must have thought their world had come apart.

The Bible provides time information about the Crucifixion. From several scriptures, we can know that the Lord was crucified on the Passover. For example, we read in John 18:39 a statement Pilate made to the crowd after the Lord Jesus had been brought to him:

> **But ye have a custom, that I should release unto you one at the passover: will ye therefore that I release unto you the King of the Jews?**

We can also determine that the day of His death was a Friday. In Mark 15:42-45, we learn that the Lord's body was put in the tomb the day before the Saturday sabbath:

> **And now when the even was come, because it was the preparation, that is, the day before the sabbath, Joseph**

> of Arimathaea, an honourable counsellor, which also waited for the kingdom of God, came, and went in boldly unto Pilate, and craved the body of Jesus. And Pilate marvelled if he were already dead: and calling *unto him* the centurion, he asked him whether he had been any while dead. And when he knew *it* of the centurion, he gave the body to Joseph.

As we continue reading this Gospel account, we learn that Joseph of Arimathaea placed the Lord's body in a sepulchre hewn out of rock, and rolled a stone to seal the entrance. Mary Magdalene and Mary who was the Lord's mother saw where the body had been laid. Very early that Sunday morning they went to the sepulchre, planning to anoint the Lord's body with spices they had brought with them. They had seen how the heavy stone sealed the entrance, and were wondering who could roll the stone away for them so that they could enter the sepulchre. When they got to the sepulchre, they saw that the stone had already been moved, as we read in Mark 16:4-7:

> **And when they looked, they saw that the stone was rolled away: for it was very great. And entering into the sepulchre, they saw a young man sitting on the right side, clothed in a long white garment; and they were affrighted. And he saith unto them, Be not affrighted: Ye seek Jesus of Nazareth, which was crucified: he is risen; he is not here: behold the place where they laid him. But go your way, tell his disciples and Peter that he goeth before you into Galilee: there shall ye see him, as he said unto you.**

Soon after this, the Lord appeared to Mary Magdalene, and then to His disciples. Eventually, hundreds of people saw the risen Lord. For forty days, the Lord was not only seen but also spoke to the apostles. Beyond any doubt, He proved to them that He was alive. After the forty days, He returned to heaven, as we read in Acts 1:9-11:

> **And when he had spoken these things, while they beheld, he was taken up; and a cloud received him out**

of their sight. And while they looked stedfastly toward heaven as he went up, behold, two men stood by them in white apparel; Which also said, Ye men of Galilee, why stand ye gazing up into heaven? this same Jesus, which is taken up from you into heaven, shall so come in like manner as ye have seen him go into heaven.

Before He was taken up, the Lord told the apostles to not depart from Jerusalem. In Acts 1:5, we read what was going to happen:

For John truly baptized with water; but ye shall be baptized with the Holy Ghost not many days hence.

In fact, the event that this verse speaks of occurred the next Pentecost. On that day, the apostles were gathered together when they experienced a sound from heaven - a sound like a rushing mighty wind. They saw tongues of fire appear and rest on them. These were tangible signs that they had been filled by the Holy Spirit. This was the baptism by the Holy Spirit about which the Lord had spoken. This day also marked the beginning of the church age, the period of time when God used local congregations to publish the Gospel and to get the Bible out all over the world.

The Crucifixion, the Resurrection and Pentecost all occurred within a period of less than three months. We know that the Crucifixion was on a Passover, and that it was a Friday. Because the day of the Passover is determined by the lunar cycle, it is possible to calculate the day on which the Passover fell in a particular year. In this way, it may be known that the year of the Crucifixion was 33 A.D. - because in that year the Passover was on a Friday. One year earlier than this, it fell on a Monday; and one year later it fell on a Wednesday. By calculation, we know that the Lord was crucified on Passover in 33 A.D. on April 1. The next day, April 2, was the Saturday sabbath. The day after that - Sunday, April 3 - was the day the Lord rose from the grave.

Earlier, we learned that Pentecost is the New Testament name for the Old Testament feast known as the Feast of Firstfruits. The date of this feast was determined by counting fifty days from the sabbath following the Passover. Counting fifty days from the April 2 Saturday sabbath brings us to Sunday, May 22. In 33 A.D. that was the day the Holy Spirit was poured out on the apostles.

There is an intriguing scripture in the Book of Daniel that ties that Pentecost day to the feast day on which the Lord was announced and on which He began His ministry. In Daniel 12:12 we read:

> **Blessed *is* he that waiteth, and cometh to the thousand three hundred and five and thirty days.**

Of course, when you read this scripture in the Book of Daniel there does not appear to be anything in it associated with the duration of the Lord's ministry. As it turns out, however, there are exactly 1,335 days from the start of His ministry (September 26, 29 A.D.) to that Pentecost day when the Holy Spirit was poured out (May 22, 33 A.D.). So this scripture in the Book of Daniel provides Biblical support for the September 26, 29 A.D. date as being the beginning of the Lord's ministry. On that date, John the Baptist announced the Lord Jesus as the Lamb of God. The pouring out of the Holy Spirit less than four years later on May 22 in 33 A.D. appears to be the last historical event in the Bible to which we can affix a date.

Before moving on, let's take a look at an amazing prophecy that points to the date of the Crucifixion. It was written over five hundred years before the Lord was even born.

A Prophecy Also Brings Us to 33 A.D.

We get confirmation of the 33 A.D. date for the Crucifixion from an amazing prophecy in the Book of Daniel. Daniel had been thinking about the seventy years of punishment mentioned in Jeremiah's prophecy, as we read in Daniel 9:1-2:

> **In the first year of Darius the son of Ahasuerus, of the seed of the Medes, which was made king over the realm of the Chaldeans; In the first year of his reign I Daniel understood by books the number of the years, whereof**

Chapter 6 God's Calendar

the word of the LORD came to Jeremiah the prophet, that he would accomplish seventy years in the desolations of Jerusalem.

(The king Darius of this scripture is actually Cyrus the Great, who conquered Babylon. We saw previously that Cyrus was a picture of the Lord Jesus.) Daniel prayed that the Lord 's anger would turn away from Jerusalem. In response to his prayer, he was visited by a messenger from heaven, identified as Gabriel. Gabriel spoke the following words to Daniel, as recorded in Daniel 9:24:

Seventy weeks are determined upon thy people and upon thy holy city, to finish the transgression, and to make an end of sins, and to make reconciliation for iniquity, and to bring in everlasting righteousness, and to seal up the vision and prophecy, and to anoint the most Holy.

Now take another look at Daniel 9:1-2. According to these verses, Daniel knew about Jeremiah's prophecy of the 70 years. He undoubtedly also knew the prophecy concerning Cyrus. Daniel had been praying about the prophecy of the seventy years, and possibly thinking of Cyrus. In answer to his prayer, God revealed to him the prophecy of the seventy weeks - a prophecy about the coming of the Lord Jesus!

The Hebrew word translated as the word "weeks" in that prophecy can also be translated as "sevens;" and so the prophecy is actually concerned with a period of 70 x 7 or 490 years required to "anoint the most Holy." The words "most Holy" must be understood as a reference to the Lord Jesus. Daniel 9:25 tells us when this period of 490 years is to begin:

Know therefore and understand, *that* from the going forth of the commandment to restore and to build Jerusalem unto the Messiah the Prince *shall be* seven weeks, and threescore and two weeks: the street shall be built again, and the wall, even in troublous times.

(This scripture actually begins another prophecy, also concerned with the Lord Jesus but using a different timeline and breaking up the seventy weeks; we are, however, applying the same starting point to the prophecy of the unbroken seventy weeks.) Note especially the words "the commandment to restore and to build Jerusalem." The issuance of this command is the starting point for the prophecy of the seventy weeks.

Let's see how this prophecy relates to an event that occurred during the Jews' time under the kings of Media-Persia. Previously, we learned that Ezra was sent back to Jerusalem in 458 B.C. to reestablish the law; but the law identifies with the Bible, which is the Gospel. The hearing of the Gospel is the means by which God saves those He has chosen. As each individual is saved, he is added to the eternal church. This eternal church - the whole body of true believers - is often compared to a building, a spiritual temple, in the Bible; but the believers are also compared to a city, as we read in Revelation 21:2:

> **And I John saw the holy city, new Jerusalem, coming down from God out of heaven, prepared as a bride adorned for her husband.**

Therefore, the seventy week prophecy of Daniel 9 tells us that the 490 years began in 458 B.C., when Ezra returned to Jerusalem to reestablish the law - which is spiritually equivalent to restoring and building Jerusalem. When we go from 458 B.C. into the future by 490 years, we come to the year 33 A.D. (remember, there is no year "0"). So we see further confirmation of that year as the date of the Crucifixion.

The End of the Historical Timeline

Much of the New Testament was written by the apostle Paul. In the Acts of the Apostles, we read about his missionary journeys in the region around the Mediterranean. All of this occurred after that Pentecost day when the Holy Spirit was poured out. The dates of events associated with those missionary journeys can only be estimated. Apparently, the last historical date that God wants us to know is May 22, 33 A.D.

Chapter 6 God's Calendar

To summarize, we have discovered the following dates for New Testament events:

Birth of the Lord Jesus	October 2, 7 B.C.
Announcement of the Lord as Lamb of God	September 26, 29 A.D.
Crucifixion	April 1, 33 A.D.
Pentecost on which Holy Spirit was poured out	May 22, 33 A.D.

Within the framework of these dates and the dates we discovered for Old Testament events, we shall see that God has embedded unmistakable numerical patterns. These patterns provide further confirmation for the dates we discovered and will actually allow us to project dates into the future to learn when end time events will occur.

Numerical Patterns in the Time Intervals

Earlier, we saw that certain numbers in the Bible have spiritual meanings, and that larger numbers may be broken down into these numbers to reveal a meaning within the number. Let us consider some of the time intervals associated with dates we discovered for Old and New Testament events.

We learned that the Old Testament was completed in 391 B.C., and that in 7 B.C. the Lord Jesus was born. The time interval between these two dates is 384 years. This number may be broken down into:

2 x 2 x 2 x 2 x 2 x 2 x 2 x 3 (end of Old Testament to birth of Lord Jesus).

Note that there are seven factors of two. The number seven is associated with the perfect fulfillment of God's purpose; the number two is associated with those who have been commissioned to bring the Gospel; the number three is associated with God's purpose. The number 384 seems to be telling us that it was God's purpose to send us the Lord Jesus as the Savior, the perfect fulfillment of God's intentions for the Gospel.

Another example associated with 391 B.C. and the life of the Lord Jesus is the time from that date to 29 A.D. inclusive. As we learned earlier, 29 A.D. was the date the Lord was publicly announced. The time interval between these dates is 419 years. If, however, we include the year 29 A.D., we get 420 years. We will see that in some instances of dealing with dates, we need to look at inclusive numbers to see a spiritual meaning. The number 420 may be broken down into:

10 x 2 x 3 x 7 (end of Old Testament to announcement of Lord Jesus, inclusive).

We see in the number 420 the same numbers (2, 3 and 7) featured by the year of the Lord's birth as it relates to 391 B.C. (the number ten is associated with completeness). Therefore, the number 420 seems to indicate God's complete purpose to fulfill His intentions for the Gospel through the Lord Jesus at the time He began His work as Messiah.

We learned that the Exodus occurred in 1447 B.C., which was the year the Lord sent plagues against Egypt - including the death of their firstborn - to deliver the children of Israel from captivity. The Bible tells us that the Lord Jesus is the firstborn of many brethren, as we read in Romans 8:28-29:

> **And we know that all things work together for good to them that love God, to them who are the called according to *his* purpose. For whom he did foreknow, he also did predestinate *to be* conformed to the image of his Son, that he might be the firstborn among many brethren.**

In 33 A.D., the Lord Jesus - identified in scripture as the firstborn - was killed for the sake of the elect. The time interval between these two dates is 1,480 years inclusive. This number breaks down to the following numbers:

Chapter 6 God's Calendar

2 x 2 x 10 x 37 (Exodus to Crucifixion, inclusive).

We know that the spiritual significance of the number 37 is God's wrath or judgment. Therefore, in the number 1,480, we see an indication that the Lord Jesus was to pay the complete penalty of God's wrath on behalf of the elect.

Here is another example of this type of pattern. This one is especially important, as we shall see later on. We learned that in the year 1877 B.C., Jacob went to Egypt during his time of great tribulation. Because of the terrible famine, he was forced to leave the land God had given him. We also learned that in 587 B.C. the nation of Judah was conquered by the Babylonians. This was a horrible time for the Jews because Jerusalem and their temple were destroyed. It was the end of their nation. This event certainly was a great tribulation. The time interval between these two dates is 1,290 years, which breaks down into:

3 x 43 x 10 (Jacob's tribulation to destruction of Jerusalem).

The number 43 has been identified with God's wrath or judgment. Therefore, the number 1,290 seems to indicate the completion of God's purpose to bring His wrath or judgment.

We have seen that many time intervals associated with Biblical dates reveal patterns because they can be factored into numbers which we have identified as having spiritual significance. There is another way in which these time intervals may reveal numbers of spiritual significance. For example, we discovered that 11,013 B.C. is the date of the creation, and that 4990 B.C. is the date of the flood. The time interval between these two dates is 6,023 years, or:

6,000 + 23 (Creation to Flood).

Recall that earlier we learned that the number 23 is associated with God's wrath or judgment.

Another such example of this type of pattern is the time interval from the flood to 33 A.D., the date of the Crucifixion. This number is 5,023 years inclusive, or:

5,000 + 23 (Flood to Crucifixion, inclusive).

In this example, as in the previous one, we see a period of time that is an integral multiple of 1,000 years added to the number 23. Perhaps God is showing us that from one date to the next there is what He considers to be a completed chapter of history as He has planned it, and that He determined to bring His judgment on the people of that day once the period had ended.

It is possible to find yet more examples of time intervals which reveal patterns showing us numbers of spiritual significance. These patterns didn't come about by chance any more than the world did. They were meticulously planned by God, and they are proof to us beyond any doubt so that we may see in them the finger of God.

Chapter 7

God's Plan

Nature reveals an incredible degree of planning. God had to know every creature's reaction to every type of situation, how well it could tolerate extremes of temperature, and how long it could go without food or water: all of these things determine the ability to survive. He had to know how a great volcanic eruption can affect global weather, and how the human body will react to every type of virus that can sicken us. He had to know if the impact of a certain sized meteorite on the earth can change the angle at which the earth tilts toward the sun. God had to plan for every contingency for all the various "systems" he created - for every kind of creature, for the weather, for the earth and for the universe - that keep nature in balance and allow life as we know it to continue. There are many billions of possible conditions. From the microscopic level all the way up to the vast scale of outer space, there is continuous movement and constant change. God's planning had to account for every possibility in order for life to exist as it does.

God Has a Salvation Plan

The Bible also reveals God as a magnificent planner; it reveals God as Creator, but it also reveals God's plan to save a people for Himself. This incredible salvation plan has been worked out in great detail, and we can learn about it as we read the Bible.

God's salvation plan is of the utmost importance. In fact, we may say that it's the reason man exists, and why man has been designed as he has been. It's the reason there is any life on earth. It is the very reason for the existence of the universe! Throughout the ages, many people have struggled for understanding about these matters. What is the purpose or meaning of life? Why were they born? Why will they die? The answer: it is God's salvation plan. How can such an "outrageous" statement possibly be supported? Remember, we're using the Bible as our authority. God's word tells us that it was God's purpose to have a people for Himself, as we read in Titus 2:13-14:

> **Looking for that blessed hope, and the glorious appearing of the great God and our Saviour Jesus Christ; Who gave himself for us, that he might redeem us from all iniquity, and purify unto himself a peculiar people, zealous of good works.**

To accomplish this purpose, He developed a plan that began before the foundation of the world. According to this plan, even the universe is expendable; but the people He chose to save will be given eternal life, as we read in Romans 6:23:

> **For the wages of sin *is* death; but the gift of God *is* eternal life through Jesus Christ our Lord.**

God's elect will be with Him for all eternity. For them, there will be no more pain, no more loneliness, no suffering of any kind. They will have regained the inheritance lost in the Garden of Eden. Unsaved mankind will be denied this inheritance. Their death, whenever it happens or will happen, will mark the end of any kind of existence for them at any time; and when the universe is destroyed, their remains will be annihilated. And this universe will be destroyed, as we read in 2 Peter 3:10:

> **But the day of the Lord will come as a thief in the night; in the which the heavens shall pass away with a great noise, and the elements shall melt with fervent heat, the earth also and the works that are therein shall be burned up.**

Chapter 7 God's Plan

This is how it will end for all unsaved persons who are still alive at the time the Lord has determined to destroy the universe, according to His plan. God's elect will have been taken off the earth before then. They will be with Him, safe from the horror that is taking place on earth.

The Lord will be coming in the near future to take the elect, both those who are alive and those who have passed away throughout all the years since creation. We read about this in 1 Thessalonians 4:15-17:

> **For this we say unto you by the word of the Lord, that we which are alive *and* remain unto the coming of the Lord shall not prevent them which are asleep. For the Lord himself shall descend from heaven with a shout, with the voice of the archangel, and with the trump of God: and the dead in Christ shall rise first: Then we which are alive *and* remain shall be caught up together with them in the clouds, to meet the Lord in the air: and so shall we ever be with the Lord.**

Every saved person who is alive at the Lord's coming will be transformed into his new, glorified body; and every saved person who has died and gone to be with the Lord in spirit will be reunited with his new, glorified body. This is further explained in 1 Corinthians 15:51-52:

> **Behold, I shew you a mystery; We shall not all sleep, but we shall all be changed, In a moment, in the twinkling of an eye, at the last trump: for the trumpet shall sound, and the dead shall be raised incorruptible, and we shall be changed.**

These elect persons, even those who were born only recently, have been known to God for many thousands of years. This is an incredible aspect of God's salvation plan, as we read in Ephesians 1:4-6:

> **According as he hath chosen us in him before the foundation of the world, that we should be holy and without blame before him in love: Having predestinated**

us unto the adoption of children by Jesus Christ to himself, according to the good pleasure of his will, To the praise of the glory of his grace, wherein he hath made us accepted in the beloved.

After the present universe is destroyed, God will create a new universe and a new earth for the elect to inhabit for all eternity. The apostle John saw this in a vision, and it is mentioned in Revelation 21:1:

And I saw a new heaven and a new earth: for the first heaven and the first earth were passed away; and there was no more sea.

It is the new earth and not this present world that is the inheritance of God's elect, as we read in Matthew 5:5:

Blessed *are* the meek: for they shall inherit the earth.

We know that God's salvation plan will end with all His elect being saved, and we know that it began before the foundation of the world. As we read the Bible, we may see distinct ages of time in which God worked throughout history to implement this plan. Just for the purpose of discussion, let's apply names to these ages as we examine some Biblical events in chronological order. We begin with the Adam-to-Noah Age.

The Adam-to-Noah Age

Early in history, beginning with Adam, there was no organization or institution to represent God's saving work on the earth. There was nothing visible - even the earliest part of the Bible was still thousands of years into the future. There were just individuals who had been already saved, or who were being drawn to God by Himself. When Adam and Eve sinned, God made a statement revealing that He had a plan. In cursing the serpent (Satan), God stated in Genesis 3:15:

And I will put enmity between thee and the woman, and between thy seed and her seed; it shall bruise thy head,

and thou shalt bruise his heel.

This statement is understood to be a prophecy of the coming of the Lord Jesus, who is the foundation of God's salvation plan. Adam and Eve may have begun looking for this seed at the time God made this statement.

In Genesis 4:1-2, we read of two brothers - Cain and Abel:

And Adam knew Eve his wife; and she conceived, and bare Cain, and said, I have gotten a man from the Lord. And she again bare his brother Abel. And Abel was a keeper of sheep, but Cain was a tiller of the ground.

Perhaps Eve thought that one of these two boys was the seed God had promised. However, Cain murdered Abel. We read in Genesis 4:8:

And Cain talked with Abel his brother: and it came to pass, when they were in the field, that Cain rose up against Abel his brother, and slew him.

As a result, the Lord drove Cain away. Genesis 4:16 states:

And Cain went out from the presence of the LORD, and dwelt in the land of Nod, on the east of Eden.

When her son Seth was born, Eve may have had in mind the Lord's statement that there would be enmity between the serpent's seed and her own. In Genesis 4:25-26, we read:

And Adam knew his wife again; and she bare a son, and called his name Seth: For God, *said she,* **hath appointed me another seed instead of Abel, whom Cain slew. And to Seth, to him also there was born a son; and he called his name Enos: then began men to call upon the name of the LORD.**

In Genesis 5, beginning with Adam's son Seth and continuing to Noah, we have a list of men who descended from Seth. Although the Bible doesn't say so, it appears that all of these men were saved by God. These names also appear in Luke 3, which traces a genealogy from Joseph - the man whom the world thought to be the father of Jesus - all the way back to Seth and Adam. In the case of Enoch, we are certain that he was one of God's elect because the Bible declares in Genesis 5:24:

> **And Enoch walked with God: and he *was* not; for God took him.**

And as for Noah, we know for certain that God saved him because of statements such as we read in Genesis 6:8:

> **But Noah found grace in the eyes of the LORD.**

The Bible also gives us the descendants of Cain in Genesis 4. We have no reason to assume that any of these individuals were ever saved. We do know that as time passed man became more wicked, so that we read in Genesis 6:5:

> **And God saw that the wickedness of man *was* great in the earth, and *that* every imagination of the thoughts of his heart *was* only evil continually.**

Because of this great wickedness, God's judgment was about to fall on the earth. Out of all mankind, God chose to save only Noah and his family from the great flood He had determined to send.

The Pre-Flood Age

The flood was one of history's greatest events. Shortly after it began, there wasn't anyone left alive other than Noah and his family. However, we are interested in the years shortly before the flood; because during this time, there was something visible - something to show God's work of salvation. That something was the ark.

The ark was a huge craft. The Bible gives us the dimensions in the units of a cubit, which we understand to be about 18 inches long. Genesis 6:15 gives us the ark's size:

> **And this *is the fashion* which thou shalt make it *of*: The length of the ark *shall be* three hundred cubits, the breadth of it fifty cubits, and the height of it thirty cubits.**

Chapter 7 God's Plan

God told Noah how to construct the ark, and undoubtedly equipped him to carry out this incredible project. Noah must have had great strength, stamina and intellect. God didn't miraculously create the ark: He had Noah build it. It would seem that only an exceptional, almost superhuman person could accomplish such a task. It must have taken a very long time to finish. How many years could it have been? In Genesis 6:3, we read:

And the LORD said, My spirit shall not always strive with man, for that he also *is* flesh: yet his days shall be an hundred and twenty years.

Although some people may believe this scripture means that man's life expectancy was going to be 120 years for a while, the succeeding life expectancies we read in Genesis are far greater than that. Therefore, it may be that in this scripture God is telling us that He gave Noah advanced notice of 120 years to build the ark. In any event, there came a time when the ark was complete enough so that it was *very* noticeable. Maybe halfway through the construction period, or a lot sooner than that, it would have gotten the attention of Noah's neighbors.

Many people surely heard about Noah. He must have been called crazy, and much worse than that. Remember, that generation had deteriorated to the point where God decided to destroy them all. For years the ark's construction went on, ever closer to completion, a sign of God's impending judgment.

From the New Testament, we learn that Noah preached. In 2 Peter 2:5, in a scripture about God's judgment on the world of Noah's day, we read:

And spared not the old world, but saved Noah the eighth *person*, a preacher of righteousness, bringing in the flood upon the world of the ungodly;

As a preacher, Noah may have warned anyone who would listen that God's judgment was coming and it was coming soon. Possibly he called on them to repent, and even told them how many years were left until the flood would come. He may have been doing that for 120 years. Not only did Noah's generation have the ark as a visible sign, they also heard him preaching to them.

When the time had almost arrived, God gave Noah the warning we read in Genesis 7:4:

> **For yet seven days, and I will cause it to rain upon the earth forty days and forty nights; and every living substance that I have made will I destroy from off the face of the earth.**

Reading this awesome warning that God gave to Noah, we should not only be thinking of what it meant to everyone who was alive at that time. When we compare this scripture with others dealing with prior occasions of God's impending judgment - for example, His judgment on Sodom and Gomorrah - we notice a pattern: God sends a warning of the time remaining before the destruction. In fact, God's word promises us such a warning; as we read in Amos 3:7:

> **Surely the Lord GOD will do nothing, but he revealeth his secret unto his servants the prophets.**

Later on, we will see how God is warning our own generation that our time is very short.

The Noah-to-Moses Age

After the flood, Noah and his descendants repopulated the earth. Genesis 11 lists descendants of Shem, Noah's son, down to Abraham (Abram). These names also appear in the list of Luke 3, so it certainly seems that all these men were saved. From many scriptures in the Bible, we know for certain that Abraham was saved. There are several chapters in the book of Genesis providing many details about Abraham and his son Isaac and Isaac's son Jacob. Whereas in one of the earlier chapters, God may have only given a name of an individual He saved, He now provides a great deal of information about these three men. I'm sure there are many spiritual lessons we can learn as we read about God's work in their lives. As we read these chapters, we learn that God established a covenant with Abraham and promised that his descendants would be innumerable. In Genesis 15:5, we read about this:

> **And he brought him forth abroad, and said, Look now toward heaven, and tell the stars, if thou be able to number them: and he said unto him, So shall thy seed be.**

From other scriptures, we now know that in this verse God was telling Abraham that He would save very many people. This wasn't just a promise that Abraham would have many descendants; no, this was a promise dealing with God's plan of salvation. This promise from Genesis is repeated in the last book of the Bible in a manner that provides further clarification for us today. In Revelation 7:9, we read:

> **After this I beheld, and, lo, a great multitude, which no man could number, of all nations, and kindreds, and people, and tongues, stood before the throne, and before the Lamb, clothed with white robes, and palms in their hands;**

This multitude of people is clothed with white robes, meaning that God has saved them and so fulfilled His promise to Abraham. These are people who are in a sense Abraham's spiritual descendants. We read about this in Galatians 3:7:

> **Know ye therefore that they which are of faith, the same are the children of Abraham.**

God also fulfilled His promise to Abraham by miraculously granting him a son, and through this son's descendants the nation of Israel eventually emerged.

Here very briefly is how it happened, according to the books of Genesis and Exodus. Abraham's son Isaac was born to him according to God's promise, despite the fact that both he and his wife Sarah were advanced in years. We read God's specific promise to them in Genesis 17:16-17:

> **And I will bless her, and give thee a son also of her: yea, I will bless her, and she shall be *a mother* of nations; kings of people shall be of her. Then Abraham fell upon**

> his face, and laughed, and said in his heart, Shall *a child* be born unto him that is an hundred years old? and shall Sarah, that is ninety years old, bear?

The promise was fulfilled, and we read in Genesis 21:5:

> And Abraham was an hundred years old, when his son Isaac was born unto him.

God continued working in the life of Isaac, and He eventually guided Abraham's servant to find the girl who became Isaac's wife, Rebekah. As she prepared to leave her family to travel to join Isaac, they blessed her and said something very interesting, as we read in Genesis 24:59-60:

> And they sent away Rebekah their sister, and her nurse, and Abraham's servant, and his men. And they blessed Rebekah, and said unto her, Thou *art* our sister, be thou *the mother* of thousands of millions, and let thy seed possess the gate of those which hate them.

The word translated as "millions" in this scripture could also have been translated as "ten thousand," a number considerably smaller than a million. Still, one thousand times ten thousand equals ten million; and so this scripture appears to be prophesying that some multiple of tens of millions of people will eventually be saved.

As we continue reading Genesis, we learn that to Isaac and Rebekah were born twin sons: Jacob and Esau. God later appeared to Isaac, even as He had appeared to Abraham; and He repeated the promise that He had made to Abraham. We read this in Genesis 26:4:

> And I will make thy seed to multiply as the stars of heaven, and will give unto thy seed all these countries; and in thy seed shall all the nations of the earth be blessed;

Through the lives of Abraham and Isaac, God was not only carrying out His salvation plan; He was also revealing it in the promises He made to them.

Just as the Bible gives us many details about Abraham and Isaac, there are also many details about Isaac's son, Jacob. In Genesis, we learn that through Jacob came the sons who eventually became the twelve tribes of ancient Israel. God appeared to Jacob and, on the occasion recorded in Genesis 35:9-11, He changed Jacob's name to Israel:

> **And God appeared unto Jacob again, when he came out of Padanaram, and blessed him. And God said unto him, Thy name *is* Jacob: thy name shall not be called any more Jacob, but Israel shall be thy name: and he called his name Israel. And God said unto him, I *am* God Almighty: be fruitful and multiply; a nation and a company of nations shall be of thee, and kings shall come out of thy loins;**

It is interesting to note that God sometimes uses the word "kings" to refer to those whom He has chosen to save, as in this promise to Jacob.

When Jacob was an old man, there was a severe famine in the land. At that time, God instructed him to go down to Egypt. This must have been a great trial for Jacob. After all, God had promised him the land where he, his father, and his father's father had been living, as we read in Genesis 35:12:

> **And the land which I gave Abraham and Isaac, to thee I will give it, and to thy seed after thee will I give the land.**

Despite the promise, Jacob is told to leave the land and go to Egypt. We read in Genesis 46:2-4:

> **And God spake unto Israel in the visions of the night, and said, Jacob, Jacob. And he said, Here *am* I. And he said, I *am* God, the God of thy father: fear not to go down into Egypt; for I will there make of thee a great nation: I will go down with thee into Egypt; and I will also surely bring thee up *again*: and Joseph shall put his hand upon thine eyes.**

At least Jacob would be consoled to see his son Joseph. Jacob had only recently learned that Joseph - who had disappeared years earlier and was presumed dead - was alive in Egypt and in a position of great power, second only to Pharaoh himself. Joseph was now asking his father to come down to Egypt, and he had even arranged to send wagons to bring the whole family there, as we read in Genesis 46:5-7:

> **And Jacob rose up from Beersheba: and the sons of Israel carried Jacob their father, and their little ones, and their wives, in the wagons which Pharaoh had sent to carry him. And they took their cattle, and their goods, which they had gotten in the land of Canaan, and came into Egypt, Jacob, and all his seed with him: His sons, and his sons' sons with him, his daughters, and his sons' daughters, and all his seed brought he with him into Egypt.**

In Genesis 46:26, we learn the size of Jacob's family at this time:

> **All the souls that came with Jacob into Egypt, which came out of his loins, besides Jacob's sons' wives, all the souls *were* threescore and six;**

From this group, God eventually created a great nation.

Because Joseph was in a position of influence, Jacob's family was cared for during the famine. In Genesis 47:11-12, we read:

> **And Joseph placed his father and his brethren, and gave them a possession in the land of Egypt, in the best of the land, in the land of Rameses, as Pharaoh had commanded. And Joseph nourished his father, and his brethren, and all his father's household, with bread, according to *their* families.**

Jacob lived to see his family flourish in the new land, and he died with his sons gathered around him. The book of Genesis ends shortly after Jacob's death with the death of Joseph.

There came a time after Joseph's death that Jacob's descendants - now greatly multiplied in number - were no longer treated well by the Egyptians. In fact, they were enslaved and treated very harshly, as we read in Exodus 1:13-14:

> **And the Egyptians made the children of Israel to serve with rigour: And they made their lives bitter with hard bondage, in mortar, and in brick, and in all manner of service in the field: all their service, wherein they made them serve, *was* with rigour.**

It was into this hostile world that Moses was born. In Exodus 2:1-2, we learn something about Moses' parents:

> **And there went a man of the house of Levi, and took *to wife* a daughter of Levi. And the woman conceived, and bare a son: and when she saw him that he *was a* goodly *child*, she hid him three months.**

In order to control their population, Pharaoh had established a policy that male babies of the Hebrews - that is, the descendants of Jacob - were to be killed. It was for this reason that the mother of Moses hid him. When she could no longer hide him, she made a little ark of bulrushes, put the baby in and placed it along the river where Pharaoh's daughter came to wash herself. Pharaoh's daughter saw the baby and must have been very taken with him because she brought him up as her son, calling him Moses.

As the adopted son of Pharaoh's own daughter, Moses must have been well educated. He would have met people who wielded power at the highest level - including Pharaoh himself - and seen how they conducted themselves. He must have enjoyed privileges that only very few in that society had. With all that he learned along the way, somehow he also learned that he was a Hebrew, not an Egyptian.

One day, he saw an Egyptian striking a Hebrew. Moses decided to kill the Egyptian. He hid the body, but his deed became known anyway. When Moses realized it, he fled to the land of Midian.

Moses made a new life for himself there and was content; but God was about to disturb his contentment. One day, God appeared to him in a burning bush. The bush burned, but was not consumed. God spoke to Moses out of the bush and told him what he had to do, as we read in Exodus 3:10:

Come now therefore, and I will send thee unto Pharaoh, that thou mayest bring forth my people the children of Israel out of Egypt.

Moses wasn't exactly enthusiastic about this assignment, to put it mildly. He just did not want to be the one to go; but God would not release him from this job. God allowed Moses to have his brother Aaron with him to confront Pharaoh, and it would be Aaron who would do the talking.

Moses alone saw the miracle of the burning bush that day. During all the time that had passed since the flood of Noah's day, there was really no outward sign of God's work of salvation. Of course, God made Himself known to at least some of the men He chose to save; in their cases there had been something for them to see or hear, but not for the world at large.

Noah, Abraham, Isaac and Jacob all built altars to God; Isaac also dug wells, and Jacob set up pillars, but we don't read that there was anything remarkable about them to distinguish them from any others built in those times. There was certainly nothing on the scale of Noah's ark. Now, at this time in Moses' life, there was about to be an outward sign of God's saving work. God Himself was about to give the world a glimpse of His power.

The Beginning of the Kingdom of Israel Until after the Crucifixion

God had raised up Moses as His instrument to deliver the children of Israel from their bondage. Through a series of plagues that devastated Egypt, God demonstrated His power. Repeatedly, Moses went to Pharaoh to demand that he let the Israelites go; each time, Pharaoh refused to let them go unconditionally - and so the next plague came upon Egypt. Finally, the time of freedom for the children of Israel was near. In Exodus 11:1, we read:

> And the LORD said unto Moses, Yet will I bring one plague *more* upon Pharaoh, and upon Egypt; afterwards he will let you go hence: when he shall let *you* go, he shall surely thrust you out hence altogether.

This last plague the Lord spoke of was to be very painful indeed for the Egyptians: it was going to be the death of all the firstborn throughout Egypt. In Exodus 12:29-32, we read:

> And it came to pass, that at midnight the LORD smote all the firstborn in the land of Egypt, from the firstborn of Pharaoh that sat on his throne unto the firstborn of the captive that *was* in the dungeon; and all the firstborn of cattle. And Pharaoh rose up in the night, he, and all his servants, and all the Egyptians; and there was a great cry in Egypt; for *there was* not a house where *there was* not one dead. And he called for Moses and Aaron by night, and said, Rise up, *and* get you forth from among my people, both ye and the children of Israel; and go, serve the LORD, as ye have said. Also take your flocks and your herds, as ye have said, and be gone; and bless me also.

And so the children of Israel were finally free. They were given not only their freedom, but valuable goods as well - as we read in Exodus 12:35-36:

> And the children of Israel did according to the word of Moses; and they borrowed of the Egyptians jewels of silver, and jewels of gold, and raiment: And the LORD gave the people favour in the sight of the Egyptians, so that they lent unto them *such things as they required.* And they spoiled the Egyptians.

Incredibly, despite all the punishment Pharaoh saw inflicted on his nation and upon his own household, he changed his mind after he had let the children of Israel go. He decided to pursue them. Pharaoh himself went after them with his army, as we read in Exodus 14:5-7:

> And it was told the king of Egypt that the people fled: and the heart of Pharaoh and of his servants was turned against the people, and they said, Why have we done this, that we have let Israel go from serving us? And he made ready his chariot, and took his people with him: And he took six hundred chosen chariots, and all the chariots of Egypt, and captains over every one of them.

During their time in Egypt, the number of Jacobs descendants increased tremendously. In Exodus 12:37, we get an indication of the number of people who constituted Israel:

> And the children of Israel journeyed from Rameses to Succoth, about six hundred thousand on foot *that were* men, beside children.

How long did it take for the 66 people who traveled with Jacob into Egypt to increase to this great number? The Bible answers this question in Exodus 12:40:

> Now the sojourning of the children of Israel, who dwelt in Egypt, *was* four hundred and thirty years.

As a result of the harsh treatment during many years of captivity, the Israelites must have been toughened. Only strong people can survive under such conditions. Yes, they were tough people. But they had no weapons, and they were not trained for war. Despite their number, they could not on their own stand against Pharaoh's army - and they knew it. In Exodus 14:9-10, we read:

> But the Egyptians pursued after them, all the horses *and* chariots of Pharaoh, and his horsemen, and his army, and overtook them encamping by the sea, beside Pihahiroth, before Baalzephon. And when Pharaoh drew nigh, the children of Israel lifted up their eyes, and, behold, the Egyptians marched after them; and they were sore afraid: and the children of Israel cried out unto the LORD.

The children of Israel had escaped from Egypt, and the LORD Himself had lead them in the wilderness. His presence was visible

Chapter 7 God's Plan

to them, day and night, as we read in Exodus 13:21-22:

> And the LORD went before them by day in a pillar of a cloud, to lead them the way; and by night in a pillar of fire, to give them light; to go by day and night: He took not away the pillar of the cloud by day, nor the pillar of fire by night, *from* before the people.

Yet now, as they were encamped by the sea, the children of Israel saw Pharaoh closing in on them. They were afraid, in spite of the visible presence of God. Only a miracle could save them from being forced back to Egypt, or even slaughtered where they stood. The Lord provided that miracle: He placed the Egyptian camp in darkness, and provided light by night to the camp of Israel. Then, in Exodus 14:21-22, we read what happened next:

> And Moses stretched out his hand over the sea; and the LORD caused the sea to go *back* by a strong east wind all that night, and made the sea dry *land*, and the waters were divided. And the children of Israel went into the midst of the sea upon the dry *ground*: and the waters *were* a wall unto them on their right hand, and on their left.

The children of Israel were able to walk right through an area that had been covered with water the day before; and this wasn't just a shallow pond, as we shall see. As they walked on dry ground through the sea, the Egyptians followed; but God didn't allow the Egyptians to pass through. In Exodus 14:27-28, we read:

> And Moses stretched forth his hand over the sea, and the sea returned to his strength when the morning appeared; and the Egyptians fled against it; and the LORD overthrew the Egyptians in the midst of the sea. And the waters returned, and covered the chariots, and the horsemen, *and* all the host of Pharaoh that came into the sea after them; there remained not so much as one of them.

With this miracle, the children of Israel were finally free of this cruel Pharaoh. He perished with his army. The Egyptians eventually found his remains on the sea shore and brought him back to Egypt for burial.

The children of Israel no longer had any reason to worry about the Egyptians. Their survival, however, was still dependent on God. Such a large number of people as they now were, in order to survive in a desert wilderness, surely required a good sized train load of food and water every day. How could they get what they needed to survive? God provided for them.

For 40 years, God led them in the wilderness. He allowed them to prevail in battle against their enemies, and so preserved their lives. By working through Moses, God brought the people close to the land He had promised. It would not be Moses, however, who would lead the children of Israel into the promised land. Moses saw the land in the distance from high ground, but the Lord chose Joshua to lead the people across the Jordan River to take the land. We read in Deuteronomy 34:4-5 the Lord's words to Moses:

> **And the LORD said unto him, This *is* the land which I sware unto Abraham, unto Isaac, and unto Jacob saying, I will give it unto thy seed: I have caused thee to see *it* with thine eyes, but thou shalt not go over thither. So Moses the servant of the LORD died there in the land of Moab, according to the word of the LORD.**

The land that Israel was about to enter, the land of Canaan, was not uninhabited. There were many hostile people there. The Bible makes it clear that Israel's success in its many battles with their armies was not due to their own superiority - it was because God gave Israel victory. God's judgment had come on the peoples who inhabited the land, and He was using the Israelites as His executioner.

The Lord picked Joshua as Moses' successor. In Deuteronomy 3:28, we read something God said to Moses about this:

> **But charge Joshua, and encourage him, and strengthen him: for he shall go over before this people, and he shall cause them to inherit the land which thou shalt see.**

Joshua led the Israelites to victory in many battles. Eventually, the day came when the land was conquered and was divided among Israel's tribes. Joshua lived to see his people go to their own inheritance in the promised land, as we read in Judges 2:6-7:

> **And when Joshua had let the people go, the children of Israel went every man unto his inheritance to possess the land. And the people served the LORD all the days of Joshua, and all the days of the elders that outlived Joshua, who had seen all the great works of the LORD, that he did for Israel.**

In the period of conquest - after Israel crossed the Jordan River and before they inherited the land - there were many dramatic miracles to attest to God's work of salvation; just as there had been during the time preceding the Exodus, and during Israel's 40 years in the wilderness. Now, their hope of many years finally realized, the people settled in their land.

God had warned the Israelites not to intermarry with any of the peoples who lived in the land, and not to worship their gods. Sadly, many of them did these things after Joshua's generation passed away; they went their own way and forsook God. In Judges 2:14, we read what then happened to the Israelites:

> **And the anger of the LORD was hot against Israel, and he delivered them into the hands of spoilers that spoiled them, and he sold them into the hands of their enemies round about, so that they could not any longer stand before their enemies.**

When the children of Israel suffered under their oppressors, they cried out to God. God would then raise up someone to deliver them from their enemies. In time, the Israelites reverted to the sins that caused God to permit their suffering. This cycle occurs repeatedly during the period chronicled in the Book of Judges. God worked miraculously during this period also. Samson, to whom God granted superhuman physical strength, is probably the best known of the deliverers God raised up during those days.

The last of these judges was named Samuel. It was during his time that the children of Israel asked for a king. In 1 Samuel 8:4-5, we read what happened:

> **Then all the elders of Israel gathered themselves together, and came to Samuel unto Ramah, And said unto him, Behold, thou art old, and thy sons walk not in thy ways: now make us a king to judge us like all the nations.**

Samuel was displeased that the people had asked for a king. He prayed to God about the matter, and God answered him. God told Samuel that the people had not rejected him - they had rejected God. Samuel was to give them their king, but he was to warn them of how the king would rule over them. This Samuel did. Despite the warning, they still wanted a king; and so he anointed Saul to be their king. Israel had become a kingdom.

Saul was king over all Israel, but it was Samuel with whom God spoke. In time, Samuel had the sad responsibility of telling Saul that the Lord had rejected him. Saul knew that God's favor now rested on David, who had killed the fearsome champion of the Philistines - Goliath. Saul tried to kill David on numerous occasions, despite David's great loyalty to him. Saul died by his own hand on the battle field, when he fell on his sword after being wounded by the Philistines. After the death of Saul, David became king over Israel.

David fought many battles in his lifetime and was a great military leader. During his reign, David made preparations for the building of a magnificent temple. David died an old man, but he lived to see his son Solomon become king, as we read in 1 Kings 2:10-12:

> **So David slept with his fathers, and was buried in the city of David. And the days that David reigned over Israel *were* forty years: seven years reigned he in Hebron, and thirty and three years reigned he in**

Chapter 7 God's Plan

Jerusalem. Then sat Solomon upon the throne of David his father; and his kingdom was established greatly.

It was during Solomon's reign that the kingdom of Israel reached its height in wealth and power. He had a substantial army, as well as a navy that appears to have carried on trade all over the region. Solomon also constructed the glorious temple for which his father David had prepared.

In the portion of the Bible telling us about Solomon's reign, we find references to other nations and their rulers. One account is the visit of the Queen of Sheba. Part of this account is given in 1 Kings 10:1-2:

> **And when the queen of Sheba heard of the fame of Solomon concerning the name of the LORD, she came to prove him with hard questions. And she came to Jerusalem with a very great train, with camels that bare spices, and very much gold, and precious stones: and when she was come to Solomon, she communed with him of all that was in her heart.**

The kingdom of Israel was a testimony to nations near and far of God's continuing salvation work. Later on in its history, Israel would have only the memory of its former greatness.

It did not take long for the decline to begin. In fact, it began during Solomon's reign, as we read in I Kings 11:4:

> **For it came to pass, when Solomon was old, *that* his wives turned away his heart after other gods: and his heart was not perfect with the LORD his God, as *was* the heart of David his father.**

Then, in 1 Kings 11:11-12, we read:

> **Wherefore the LORD said unto Solomon, Forasmuch as this is done of thee, and thou hast not kept my covenant and my statutes, which I have commanded thee, I will surely rend the kingdom from thee, and will give it to thy servant. Notwithstanding in thy days I will not do it for David thy father's sake: *but* I will rend it out of the**

hand of thy son.

Even before the death of Solomon, God raised up enemies against him. When Solomon died, his son Rehoboam succeeded him as king. Soon after his reign began, however, the main part of Rehoboam's kingdom broke away under a man named Jeroboam. From then on, the part that broke away - consisting of ten tribes - came to be known as the Kingdom of Israel, or the Northern Kingdom. Only Judah, which was David's own tribe, and the tribe of Benjamin followed Rehoboam; this part came to be known as the Kingdom of Judah, or the Southern Kingdom.

Sadly, from what we read about the kings of both kingdoms, it appears that most of them were not saved. At times, their behavior was even worse than that of the nations God had destroyed by the hand of Israel, as we read in 2 Chronicles 33:9:

> **So Manasseh made Judah and the inhabitants of Jerusalem to err, *and* to do worse than the heathen, whom the LORD had destroyed before the children of Israel.**

God sent prophets to each of the two kingdoms throughout their history to warn them of their sins. Many of the prophets suffered persecution and even death in order to bring the messages they received from God. In time, God caused both kingdoms to be conquered.

The first of the two kingdoms to be conquered was the Kingdom of Israel. This occurred in 709 B.C., when it was conquered by the Assyrians. (Most historians believe it happened in 722 B.C., but the Biblical evidence doesn't support that date.) The Kingdom of Judah persisted for another 122 years until, in 587 B.C., it too was conquered. It was the Babylonians, under Nebuchadnezzar the king of Babylon, who conquered the Kingdom of Judah.

Even in captivity, God's eyes were on His people to preserve them. They retained their identity, as we see in Psalm 137:1-5:

Chapter 7 God's Plan

> By the rivers of Babylon, there we sat down, yea, we wept, when we remembered Zion. We hanged our harps upon the willows in the midst thereof. For there they that carried us away captive required of us a song; and they that wasted us *required of us* mirth, *saying*, Sing us *one* of the songs of Zion. How shall we sing the LORD'S song in a strange land? If I forget thee, O Jerusalem, let my right hand forget *her cunning*.

Time passed, and the nation that had conquered the Kingdom of Judah was itself conquered. This occurred in 539 B.C. when Cyrus, king of the Media-Persian empire, conquered Babylon. Shortly thereafter, something incredible happened: Cyrus allowed the Jews - who were now living in Cyrus' greatly expanded kingdom - to return to Jerusalem. In a sense, Cyrus' decree allowing the Jews to return was not really surprising. This is so because God had announced nearly 200 years earlier that there would be a man named Cyrus who would do this very thing. We read about it in Isaiah 44:28, which is another one of the Bible's amazing prophecies that has been fulfilled in history:

> **That saith of Cyrus, *He is* my shepherd, and shall perform all my pleasure: even saying to Jerusalem, Thou shalt be built; and to the temple, Thy foundation shall be laid.**

When Nebuchadnezzar conquered the Kingdom of Judah, he destroyed the temple built during the reign of Solomon, hundreds of years earlier. Now, in 539 B.C., the returning Jews could begin rebuilding the temple. It should be noted that this rebuilt temple was very different than the one the Lord Jesus saw and walked in a few hundred years later. In John 2:19-20, we read:

> **Jesus answered and said unto them, Destroy this temple, and in three days I will raise it up. Then said the Jews, Forty and six years was this temple in building, and wilt thou rear it up in three days?**

The rebuilding that began in 539 B.C. continued for many years. Even at its most complete or best stage, this temple must have been very modest compared to the one constructed in Solomon's time All the same, it was the focal point of Jewish life.

Israel continued to exist as a nation - with a temple - right up to the time of the Lord Jesus, although by then they were under the iron hand of Rome. They continued to be the visible evidence of God's continual saving work.

We may say that the birth of the Lord Jesus to a virgin named Mary in 7 B.C. is the pivotal event in God's salvation plan. A miraculous star appeared at the time of His birth. We know that the star was miraculous and very low in the sky because it pinpointed the exact location where the Lord was. Only a miraculous star could lead anyone to a particular building. Around that time there were other miraculous events as well; but in the town of Nazareth where the Lord Jesus now lived, it was only Mary and her husband Joseph who recognized the Lord Jesus as being special, so far as we can tell. The situation changed when the Lord Jesus began His ministry.

During the time of His ministry - it lasted about three and a half years - He performed many spectacular miracles and was considered by many Jews to be a prophet; but relatively few recognized Him as God. Ultimately, He was rejected by the Jewish nation and died a horrible death. I think it is safe to say that, so far as the world at that time saw it, His life was of little or no consequence. Except for the relatively small number of people whose lives were forever changed by one of His healing miracles or by His saving grace, His life didn't yet mean anything to the world at large.

This continued to be the situation even after His resurrection. The Lord Jesus was seen by hundreds of people from the Sunday after His death until the time He was taken up to heaven 40 days later. The Lord's disciples remembered how, before His death, He had said "Destroy this temple, and in three days I will raise it up." They came to understand that the Lord had been speaking of His own body, and so they believed the scriptures, as we read in John 2:22:

When therefore he was risen from the dead, his disciples remembered that he had said this unto them; and they believed the scripture, and the word which Jesus had said.

The Jewish leaders, instead of seeing the hand of God in the empty tomb, saw only a sect they considered to be a growing problem.

The Jewish nation continued to exist until 70 A.D., when they were destroyed by the Romans. Even after this catastrophe, the Jews maintained their identity down through all the centuries, and once again became a nation in the year 1948; but does that mean they continued to represent God's continuing salvation plan? No, it doesn't. God was no longer with them as He had been, as we shall see.

Even before they became a nation, when they were in the wilderness after leaving Egypt, God was with them. By day, He revealed His presence as a pillar of fire, and by night as a pillar of cloud. God did many mighty miracles for the nation of Israel, and demonstrated His power and presence dramatically many times throughout their history. One occasion was the dedication of the temple, during Solomon's reign, as we read in 2 Chronicles 7:1-3:

Now when Solomon had made an end of praying, the fire came down from heaven, and consumed the burnt offering and the sacrifices; and the glory of the LORD filled the house. And the priests could not enter into the house of the LORD, because the glory of the LORD had filled the LORD'S house. And when all the children of Israel saw how the fire came down, and the glory of the LORD upon the house, they bowed themselves with their faces to the ground upon the pavement, and worshipped, and praised the LORD, *saying*, For *he is* good; for his mercy *endureth* for ever.

Now that the Lord Jesus had come, God's relationship to this people would change. He would no longer be with the Jewish nation as He had been for so many centuries. How can we know this? We know that God is no longer working with the Jewish people as a whole because they don't recognize the Lord Jesus as God, and it is only through Him that any one may be saved. The apostle Peter explained this to the Jewish high priest and others, referring to a man who had been miraculously healed, as we read in Acts 4:10-12:

> **Be it known unto you all, and to all the people of Israel, that by the name of Jesus Christ of Nazareth, whom ye crucified, whom God raised from the dead, *even* by him doth this man stand here before you whole. This is the stone which was set at nought of you builders, which is become the head of the corner. Neither is there salvation in any other: for there is none other name under heaven given among men, whereby we must be saved.**

To help us understand what would happen to the nation of Israel, we also have in the scriptures a parable that the Lord Jesus told. In Luke 20:9-16, we read:

> **Then began he to speak to the people this parable; A certain man planted a vineyard, and let it forth to husbandmen, and went into a far country for a long time. And at the season he sent a servant to the husbandmen, that they should give him of the fruit of the vineyard: but the husbandmen beat him, and sent *him* away empty. And again he sent another servant: and they beat him also, and entreated *him* shamefully, and sent *him* away empty. And again he sent a third: and they wounded him also, and cast *him* out. Then said the lord of the vineyard, What shall I do? I will send my beloved son: it may be they will reverence *him* when they see him. But when the husbandmen saw him, they reasoned among themselves, saying, This is the heir: come, let us kill him, that the inheritance may be ours. So they cast him out of the vineyard, and killed *him*. What therefore shall the lord of the vineyard do unto them? He shall come and destroy these husbandmen, and shall give the vineyard to others. And when they heard *it*, they said, God forbid.**

Chapter 7 God's Plan

Many times, God had sent prophets to warn His people of their sins. These faithful servants of God were often mistreated, like the servants of the man who owned the vineyard. Finally, God sent His son, in the person of the Lord Jesus; and like the son of the man who owned the vineyard, the Son of God was killed. Even as early as the time of Abraham, God's word pointed to the Lord Jesus. In Genesis 22:7-8, we read:

> **And Isaac spake unto Abraham his father, and said, My father: and he said, Here *am* I, my son. And he said, Behold the fire and the wood: but where *is* the lamb for a burnt offering? And Abraham said, My son, God will provide himself a lamb for a burnt offering: so they went both of them together.**

The Jews, who considered themselves sons of Abraham, just couldn't understand that the Saviour to come would be the Lamb of God; and so they rejected the Lord Jesus. The time was coming when God would be finished using the Jewish nation - the vineyard would be given to others. But when did this happen?

The Gospel accounts tell us what happened when the Lord Jesus died on the Passover. In Matthew 27:50-51, we read:

> **Jesus, when he had cried again with a loud voice, yielded up the ghost. And, behold, the veil of the temple was rent in twain from the top to the bottom; and the earth did quake, and the rocks rent;**

The veil of the temple kept the most holy part of it in darkness. Only once a year was the high priest allowed to go past the veil, to enter that part of the temple. In 1 Kings 8:12-13, we read:

> **Then spake Solomon, The LORD said that he would dwell in the thick darkness. I have surely built thee an house to dwell in, a settled place for thee to abide in for ever.**

Once the veil was torn, that part of the temple was no longer in thick darkness. Perhaps it was at the moment God tore the veil that He made it official: God would no longer be with the Jews as He had been, and national Israel would no longer represent the eternal kingdom of God. God's salvation plan was about to enter a new age.

Pentecost and the Church Age

After His resurrection, the Lord Jesus commanded the apostles to wait in Jerusalem, as we read in Acts 1:4:

> **And, being assembled together with *them*, commanded them that they should not depart from Jerusalem, but wait for the promise of the Father, which, *saith he*, ye have heard of me.**

In Acts 1:8, the Lord Jesus provided further information about what the apostles were to wait for:

> **But ye shall receive power, after that the Holy Ghost is come upon you: and ye shall be witnesses unto me both in Jerusalem, and in all Judaea, and in Samaria, and unto the uttermost part of the earth.**

This scripture was fulfilled less than two months later. On the Pentecost following the Passover on which the Lord Jesus had been crucified, the apostles received the Holy Spirit. In Acts 2:1-4, we read:

> **And when the day of Pentecost was fully come, they were all with one accord in one place. And suddenly there came a sound from heaven as of a rushing mighty wind, and it filled all the house where they were sitting. And there appeared unto them cloven tongues like as of fire, and it sat upon each of them. And they were all filled with the Holy Ghost, and began to speak with other tongues, as the Spirit gave them utterance.**

When the apostles spoke in foreign languages they couldn't possibly have learned in any normal way, the news of it spread. As it happened - and of course none of these things just "happen" because God is directing all of it - there were at that time in

Jerusalem many Jews from different nations. These men heard the news and came to witness the strange phenomenon involving the apostles. What they heard when they came was the word of God being preached in their native tongue. They heard the Gospel. There was amazement among those who had come; but some mocked the apostles, implying that they were drunk. Then the apostle Peter addressed the group, explaining to them about the death and resurrection of the Lord Jesus. In Acts 2:37-38, we read:

> **Now when they heard *this*, they were pricked in their heart, and said unto Peter and to the rest of the apostles, Men *and* brethren, what shall we do? Then Peter said unto them, Repent, and be baptized every one of you in the name of Jesus Christ for the remission of sins, and ye shall receive the gift of the Holy Ghost.**

Among those who heard Peter that day were many of God's elect, because in Acts 2:41-42 we read:

> **Then they that gladly received his word were baptized: and the same day there were added *unto them* about three thousand souls. And they continued stedfastly in the apostles' doctrine and fellowship, and in breaking of bread, and in prayers.**

In Acts 2:47, we read this about the believers:

> **Praising God, and having favour with all the people. And the Lord added to the church daily such as should be saved.**

The Holy Spirit had been poured out, and the church age had begun. For hundreds of years, God had used the children of Israel, then the unified Kingdom of Israel, then the divided kingdoms (the Kingdom of Israel and the Kingdom of Judah) and finally the Jewish nation to represent God's elect, the eternal kingdom of God. Now in the church age He would use the earthly church to represent the eternal church - those who are God's elect.

On that Pentecost day, God's salvation plan had taken a dramatic new turn as He saved three thousand people. The church at Jerusalem was off to a spectacular start. A short time later another large group of people believed the Gospel after seeing the result of a miraculous healing. There was a lame man who was carried daily to a spot outside the temple to ask for alms. When the people saw him walking, leaping and praising God, and recognized him as the man who had been sitting at the temple gate, they ran together, greatly wondering. Once again, Peter addressed the crowd. In Acts 4:4, we read the result:

> **Howbeit many of them which heard the word believed and the number of the men was about five thousand.**

Before long, however, a persecution arose. It began when Stephen was martyred. Stephen was one of seven men chosen to assist the apostles. God used him in an exceptional way, as we read in Acts 6:8:

> **And Stephen, full of faith and power, did great wonders and miracles among the people.**

Stephen was falsely accused of blasphemy as a result of his discussions with members of a certain synagogue. He was seized and brought before the religious leaders. The high priest gave him an opportunity to speak and Stephen did so, delivering a substantial sermon. What Stephen said infuriated the religious leaders, as we read in Acts 7:57-58:

> **Then they cried out with a loud voice, and stopped their ears, and ran upon him with one accord, And cast *him* out of the city, and stoned *him*: and the witnesses laid down their clothes at a young man's feet, whose name was Saul.**

Stephen's sermon resulted in his death and appears to have triggered the persecution we read about in Acts 8:1:

> **And Saul was consenting unto his death. And at that time there was a great persecution against the church which was at Jerusalem; and they were all scattered abroad throughout the regions of Judaea and Samaria, except the apostles.**

This Saul (not to be confused with the Saul who was Israel's first king more than a thousand years earlier) later became the apostle Paul; but at this time in his life, he was busy persecuting the church. God apparently used this persecution - and Paul himself may have been the architect of it and its chief instrument - to spread the Gospel.

We know that before the persecution began, the church at Jerusalem had grown substantially. We read in Acts 6:7:

And the word of God increased; and the number of the disciples multiplied in Jerusalem greatly; and a great company of the priests were obedient to the faith.

Now many of these disciples, among them a large number of Jewish priests who were familiar with the Old Testament scriptures, were fleeing persecution. They would spread the Gospel throughout the region. Eventually it would reach throughout the Roman Empire and beyond.

From the beginning of the church age on Pentecost in 33 A.D. until the present, almost 2,000 years have elapsed. Today, roughly a third of the world's population - about two billion people - consider themselves to be Christians. Many of them are church-going people who belong to a local congregation, and so there are very many of these congregations all over the world. Throughout the church age, the local congregations have been the visible evidence of God's saving work and representative of God's eternal church; but does it continue that way until the Lord's return, or does there come a time when the church age ends? Will God ever finish His saving work in the local congregations, as He finished using the Jewish nation after the Crucifixion?

The Wheat and the Tares

The Bible provides answers to these questions. First, we must understand that not everyone who identifies himself as a Christian is one of God's elect - a true believer. As early as the first century, the churches were being infiltrated by people who were not true believers. We know this is so from certain scriptures in the book of Revelation. Chapters 2 and 3 of this amazing book include messages to seven churches. In Revelation 2:7, we read:

> **He that hath an ear, let him hear what the Spirit saith unto the churches; To him that overcometh will I give to eat of the tree of life, which is in the midst of the paradise of God.**

These seven churches give us a picture of earthly churches throughout the church age. As we read the messages, we see that almost from the start there were serious problems in the churches. In Revelation 2:14, we read part of the message to the church in Pergamos:

> **But I have a few things against thee, because thou hast there them that hold the doctrine of Balaam, who taught Balac to cast a stumblingblock before the children of Israel, to eat things sacrificed unto idols, and to commit fornication.**

Many of the people in these early churches were not saved: they were the tares - that is, weeds - the Lord Jesus spoke of in His parable of the tares in the field. In this parable, He likened the kingdom of heaven to a man who sowed good seed in his field. While men slept, his enemy came and sowed weeds among the wheat, and went his way. In time, the weeds appeared. The remainder of the parable is in Matthew 13:27-30:

> **So the servants of the householder came and said unto him, Sir, didst not thou sow good seed in thy field? from whence then hath it tares? He said unto them, An enemy hath done this. The servants said unto him, Wilt thou then that we go and gather them up? But he said, Nay; lest while ye gather up the tares, ye root up also the wheat with them. Let both grow together until the harvest: and in the time of harvest I will say to the reapers, Gather ye together first the tares, and bind them in bundles to burn them: but gather the wheat into my barn.**

As one might expect, those who identify themselves as Christians have usually been members of a local congregation; but this is where the tares are - they are in local congregations all over the world. They aren't in mosques, Buddhist temples or Hindu shrines: they are in Christian churches.

Only God knows who is a tare and who is one of His own. People who see the tares cannot tell the difference because tares can imitate true believers. In fact, many of these tares actually *believe* they have become saved. They are trusting in their own decision to become a Christian ("to accept Christ") and trusting in their local congregations - in their churches - for their salvation. They faithfully attend church services and do whatever else is expected of church members. By doing these things, they believe they can be saved; but salvation isn't a matter of what we decide, or what we do, or our own work. It is totally dependent on God.

God lets the tares coexist with the true believers until the time He decides to gather them. Notice that the tares are gathered first: "Gather ye together first the tares, and bind them in bundles to burn them: but gather the wheat into my barn." This occurs when God's judgment process begins. In 1 Peter 4:17, we read:

For the time *is come* that judgment must begin at the house of God: and if *it* first *begin* at us, what shall the end *be* of them that obey not the gospel of God?

At first glance, this scripture seems to be dealing only with the apostle Peter's day; but this scripture is certainly applicable to our day. There is a judgment period that begins at the house of God. And what is the house of God? It is the earthly church, the local congregations all over the world. This is where the tares are being gathered because the Bible commands true believers to leave the local congregations!

God Finishes With the Earthly Church Near the End of Time

How can it be that the Bible commands true believers to leave the local congregations? Why would God ever do such a thing, and where is such a command stated in the Bible? We should remember that the Lord Jesus had some very harsh things to say about the religious leaders of His day, as the time neared when God would finish using the nation of Israel. In Matthew 23:27, He said:

> **Woe unto you, scribes and Pharisees, hypocrites! for ye are like unto whited sepulchres, which indeed appear beautiful outward, but are within full of dead *men's* bones, and of all uncleanness.**

Incredibly, in a letter written to the church at the city of Corinth after the church age had begun, we find this warning in 2 Corinthians 11:13-15:

> **For such *are* false apostles, deceitful workers, transforming themselves into the apostles of Christ. And no marvel; for Satan himself is transformed into an angel of light. Therefore *it is* no great thing if his ministers also be transformed as the ministers of righteousness; whose end shall be according to their works.**

These false apostles, deceitful workers, are church leaders who appear to be ministers of righteousness but who are actually ministers of Satan! Just as the early church was being "seeded" with tares - people who looked like true believers but were not - the early church was also being infiltrated by false apostles. Now you may be wondering if this situation which occurred ages ago may have been corrected, so that the church could go on to a glorious future until the end of time. Sadly, that is not the case. As we read the Bible, we find that indeed the situation did develop ages ago - it developed almost as soon as the church age began. As the centuries rolled by, however, the situation continued to deteriorate. We learn that this is so by reading 2 Timothy 3:1-5:

> **This know also, that in the last days perilous times shall come. For men shall be lovers of their own selves, covetous, boasters, proud, blasphemers, disobedient to parents, unthankful, unholy, Without natural affection,**

> trucebreakers, false accusers, incontinent, fierce, despisers of those that are good, Traitors, heady, highminded, lovers of pleasures more than lovers of God; Having a form of godliness, but denying the power thereof: from such turn away.

Notice that this is a warning about the last days and men who will have a form of godliness. These are the false apostles of our own age. Another warning about the last days is found in 2 Thessalonians 2:1-4:

> **Now we beseech you, brethren, by the coming of our Lord Jesus Christ, and *by* our gathering together unto him, That ye be not soon shaken in mind, or be troubled, neither by spirit, nor by word, nor by letter as from us, as that the day of Christ is at hand. Let no man deceive you by any means: for *that day shall not come*, except there come a falling away first, and that man of sin be revealed, the son of perdition; Who opposeth and exalteth himself above all that is called God, or that is worshipped; so that he as God sitteth in the temple of God, shewing himself that he is God.**

The "day of Christ" will not arrive until the "son of perdition" is sitting in the temple of God. If we believe that all of these scriptures are indeed the words of God, we cannot dismiss this warning. God is telling us that as we get close to the end, Satan will be ruling in the church! True believers are to leave the church when this time comes.

We get an indication of when this time will come from the words of the Lord Jesus recorded in Matthew 24:11-14:

> **And many false prophets shall rise, and shall deceive many. And because iniquity shall abound, the love of many shall wax cold. But he that shall endure unto the end, the same shall be saved. And this gospel of the kingdom shall be preached in all the world for a witness unto all nations; and then shall the end come.**

Notice that the end doesn't come until the gospel of the kingdom is preached in all the world. This is happening today, and more so than at any other time in history. It wasn't until the final years of the last century that technology made it all possible. Now, dozens of satellites in orbit around the earth, shortwave radio and the Internet send the gospel out to the whole world. Even today, there are places where you will risk your life by carrying a Bible; but modern technology can reach the people there.

Continuing with the Lord's words in Matthew 24:15-16, we read:

When ye therefore shall see the abomination of desolation, spoken of by Daniel the prophet, stand in the holy place, (whoso readeth, let him understand:) Then let them which be in Judaea flee into the mountains:

Here again we see a reference to the time when Satan (the abomination of desolation) is in the church (the holy place). We already know that this happens some time near the end of the world - it cannot be something that happened hundreds or thousands of years ago. When we see this, the Lord commands those "in Judaea" to "flee into the mountains." That is, God's elect are to leave the church at that time.

Incredibly, even in many Old Testament scriptures we find veiled references to the end of the church age. Read, for example Isaiah 4:1:

And in that day seven women shall take hold of one man, saying, We will eat our own bread, and wear our own apparel: only let us be called by thy name, to take away our reproach.

The seven women want to eat their own bread; that is, they want their own gospel, which is a false gospel. They want to wear their own apparel: that is, they want to be clothed in their own works-based righteousness rather than in the righteousness of the Lord Jesus. Finally, they want to be called by the name of the "one man" - they want to be called Christians. These seven women must represent the local congregations after God has finished using them.

Revelation 18 tells us about one of the visions God gave to the apostle John after he had been exiled to the island of Patmos. In that chapter, all of the local congregations are represented by a single woman. Read Revelation 18:1-3:

> **And after these things I saw another angel come down from heaven, having great power; and the earth was lightened with his glory. And he cried mightily with a strong voice, saying, Babylon the great is fallen, is fallen, and is become the habitation of devils, and the hold of every foul spirit, and a cage of every unclean and hateful bird. For all nations have drunk of the wine of the wrath of her fornication, and the kings of the earth have committed fornication with her, and the merchants of the earth are waxed rich through the abundance of her delicacies.**

You may recall that Revelation chapters 2 and 3 tell us about seven churches (recall the seven women who want to eat their own bread and wear their own apparel). These churches are a picture of all the churches throughout the church age. Revelation 18 shows us what those churches have become and why God is finished with them.

The Number of the Beast and the End of the Church Age

Many have puzzled over the meaning of the "number of the beast" found in Revelation 13:18, where we read:

> **Here is wisdom. Let him that hath understanding count the number of the beast: for it is the number of a man; and his number *is* Six hundred threescore *and* six.**

As we read Revelation 13, we must conclude that the beast is identified with Satan. This conclusion is also supported by what we read in 2 Thessalonians 2:3:

> **Let no man deceive you by any means: for *that day shall not come*, except there come a falling away first, and that man of sin be revealed, the son of perdition;**

Compare that scripture with Isaiah 14:16:

> **They that see thee shall narrowly look upon thee, and consider thee,** *saying,* **Is this the man that made the earth to tremble, that did shake kingdoms;**

As we read Isaiah 14, we find that it is clearly telling us about Satan. Notice that Satan is called a man in the preceding two scriptures, and the number of the beast is the number of a man. Therefore, the number of the beast is a number God is applying to Satan.

What could be the spiritual message behind this number? Notice that 666 breaks down into the following significant factors: 2 x 3 x 3 x 37. Based on what we have learned of God's use of numbers, we can say that the message embedded in the number of the beast is the following: it is God's purpose (the number 3) to bring His wrath or judgment (the number 37) against those commissioned to bring the Gospel (the number two). Of course, as we examine this numerical pattern we need to remember that Satan and his ministers also bring the Gospel; but theirs is a false gospel.

We've seen similar numerical patterns coming from some of the time intervals we analyzed; but the number 666 is identified with Satan himself, and Satan in turn is being identified with those whom God has commissioned to bring the Gospel (via the number two as a factor in the number of the beast). Therefore, in the number 666 we see further confirmation of what so many scriptures are telling us: God has ended the church age and installed Satan to rule there.

A Scripture We Need To Examine Carefully: Hebrews 10:25

A scripture that poses difficulty for people who are struggling to believe that God could ever be finished with the local congregations is Hebrews 10:25. The problem is that this verse seems to imply that the church age will never end. Let's take a closer look at this verse and the verse preceding it. In Hebrews 10:24-25 we read:

> **And let us consider one another to provoke unto love and to good works: Not forsaking the assembling of ourselves together, as the manner of some *is*; but exhorting *one another:* and so much the more, as ye see the day approaching.**

When someone reads these two verses, it's no surprise if he or she assumes they apply to the physical church. The phrase "assembling of ourselves together" especially seems to point to a church meeting. If we look at some of the key words in the original Greek, however, a different picture begins to emerge. For example, the word "provoke" in verse 24 is the Greek word "paroxusmos" (Strong's number 3948, pronounced *par-ox-oos-mos'*). We find this same word used in Acts 15:39, which reads:

> **And the contention was so sharp between them, that they departed asunder one from the other: and so Barnabas took Mark, and sailed unto Cyprus;**

In Acts 15:39, that same Greek word has been translated as "contention." If you check Strong's concordance for the word "sharp" to see which Greek word was translated as "sharp," you find that there is no number listed next to the entry for its use in Acts 15:39. The reason is that the meaning for the word "sharp" is incorporated into the Greek word "paroxusmos." In fact, when we read Acts 15, we find that Paul and Barnabas, two children of God who had worked and traveled together extensively and by whom God had worked miracles and wonders, had such a sharp disagreement over a matter that they parted company over it. So the Greek word used to describe this situation with Paul and Barnabas is the same word God used in Hebrews 10:24, where we are instructed to consider one another "to provoke unto love and to good works." (Although you will find a number of scriptures that use either of the two English words "contention" or "provoke" throughout the New Testament, the Greek word "paroxusmos" is translated these ways only in Acts 15:39 and Hebrews 10:24). There is certainly similarity between our own situation and the disagreement between these two men when we realize that the local congregations have "fallen away" from the truth and that we are commanded to leave.

Now, let's consider Hebrews 10:25. (Notice the reference to the day of judgment in the words "as ye see the day approaching.") When we check the concordance for the word "assembling" to try to understand what is meant by the phrase "Not forsaking the assembling of ourselves together," we find that it is the Greek word "episunagoge" (Strong's number 1997, pronounced *ep-ee-soon-ag-o-gay'*). That Greek word is derived from another Greek word: "episunago" (Strong's number 1996, pronounced *ep-ee-soon-ag'-o*) which according to the concordance - means "gather (together)." There is really no separate Greek word in this scripture that means "together." The "togetherness" is included in the meaning. Of course, a word's meaning according to a concordance or other Bible reference material may be helpful; but we need to remember that the Bible is its own dictionary. The way God uses a word guides us to understand its meaning.

To further help us understand, let's take a look at 2 Thessalonians 2:1. There, we read:

Now we beseech you, brethren, by the coming of our Lord Jesus Christ, and *by* our gathering together unto him,

The words "gathering together" are the translation from the Greek word "episunagoge." It is, in this scripture, a gathering together to meet the Lord Jesus Christ at His coming. (Note that it is only in 2 Thessalonians 2:1 that "episunagoge" is used to mean "gathering," and it is only in Hebrews 10:25 that it is used to mean "assembling.") Therefore, the indication is that the assembling of ourselves, according to Hebrews 10:25, is an assembling before God in prayer. Perhaps God sees His elect this way - assembled before Him. We come before Him in prayer one by one from all over the earth, and that is how He will take us to be with Him at His coming, as we read in Matthew 24:40-42:

Then shall two be in the field; the one shall be taken, and the other left. Two *women shall be* grinding at the mill; the one shall be taken, and the other left. Watch

Chapter 7 God's Plan

therefore: for ye know not what hour your Lord doth come.

Finally, let's consider the phrase "exhorting one another" from Hebrews 10:25. Right away, we notice that the words "one another" are in italics. The translators added those words because they thought they helped to clarify the meaning, but those words are not part of the original inspired text. The Greek word translated as "exhorting" is "parakaleo" (Strong's number 3870, pronounced *par-ak-al-eh'-o*). To help us understand what may be meant by the word "exhorting" as it is used in Hebrews 10:25, let's take a look at another scripture where it's used. In 2 Corinthians 1:4, we read:

Who comforteth us in all our tribulation, that we may be able to comfort them which are in any trouble, by the comfort wherewith we ourselves are comforted of God.

The Greek word translated as "comforteth" is "parakaleo," which is also the word that has been translated as the verb "comfort" in this verse ("to comfort"). It is in God that we receive our comfort. Looking now at our original scripture in the light of 2 Corinthians 1:4, we may understand Hebrews 10:25 to be telling us that we - each one of us - should be comforted in God, "and so much the more, as ye see the day approaching." Therefore, we see that the meaning of Hebrews 10:25 is not inconsistent with our understanding that the church age has come to an end.

A Shipwreck and the End of the Church Age

There is an interesting incident about a shipwreck recorded in the Acts of the Apostles. The apostle Paul had been arrested as a result of his missionary work. He was being taken to Rome, under Roman guard, to stand trial. While at sea, the ship was caught in a fierce tempest that drove it for days. Eventually, the crew managed to get the ship close enough to shore so that crew and passengers could make it to land. We read about this in Acts 27:41:

And falling into a place where two seas met, they ran the ship aground; and the forepart stuck fast, and

remained unmoveable, but the hinder part was broken with the violence of the waves.

Some were able to swim to shore; others made it to shore by clinging to pieces of the wreckage. Although the ship was a total loss, there was no loss of life, as we read in Acts 27:44:

And the rest, some on boards, and some on *broken pieces* of the ship. And so it came to pass, that they escaped all safe to land.

As we read the Bible, we learn that the destroyed ship represents the end of the Church Age. The number of those aboard that ship - 276, as we read in Acts 27:37 - is especially significant:

And we were in all in the ship two hundred threescore and sixteen souls.

Once again, we need to be aware of the significance of numbers in the Bible and the idea of breaking down a number into significant factors. Let's apply this to the number 276. We find that it can be broken down as follows: 3 x 2 x 2 x 23. Through its significant factors, we may interpret the number 276 as telling us that it is God's purpose (3) that those who are to continue to bring the true Gospel (2) would be rescued from the judgment (23) God will bring on the local congregations (because they were saved from the doomed ship).

There are other scriptures, in both the Old and the New Testament, that tell us that God would eventually finish using the local congregations. As we take a look at the annual feasts, we will find further confirmation of this.

The Annual Feast Days and the Salvation Plan

We previously learned that we are no longer required to keep the annual feasts given to ancient Israel. That does not mean we should disregard what the Bible tells us about them. In fact, these annual feasts provide us with insight into God's salvation plan. Many Christians know this. There is, however, an assumption that many make when trying to correlate these feasts with events in God's plan; this is the assumption that the chronological progression of each feast day, as it occurs in the lunar calendar from the first month to the seventh month, corresponds to an event in God's salvation plan as it has occurred and will occur chronologically. We cannot make this assumption or any other assumption unless we are guided by the Bible to do so.

The Passover - the first feast listed in Leviticus chapter 23 - occurred in the first month, on the fourteenth day; the Feast of Tabernacles, the last feast listed in that chapter, began in the seventh month on the fifteenth day. In between are the other feasts, listed chronologically.

When we see how Passover and Pentecost were fulfilled in 33 A.D. in the New Testament, we can easily make the assumption that the remaining feasts - in the order in which they are listed - point to salvation plan events that have occurred or will occur in chronological order relative to these two days; but what if this is an incorrect assumption?

The feast known as the Day of Atonement occurred in the seventh month on the tenth day. It was on this day, and only on this one day every year, that the high priest was permitted to enter the most holy area of the temple. Before he could pass through the veil to enter this area, the Holy of Holies, he had to make elaborate preparations. These preparations included the sprinkling of blood, symbolizing atonement for the sins of all Israel. After passing the veil, the high priest was in close proximity to the ark - the very spot where God had told Moses He would appear, as we read in Leviticus 16:2:

> **And the LORD said unto Moses, Speak unto Aaron thy brother, that he come not at all times into the holy *place* within the veil before the mercy seat, which *is* upon the ark; that he die not: for I will appear in the cloud upon the mercy seat.**

We have seen that there is much Biblical evidence to indicate that it was on the Day of Atonement in the year 7 B.C. that the Lord Jesus was born. Just as the high priest of Israel was permitted on this one day a year to be near a physical appearance of God Himself within the temple, the Lord Jesus came on this day to dwell with men.

The feast known as the Feast of Trumpets could be called the Feast of Jubilee. It occurred in the seventh month, on the first day. On that day, the shophar, a hollowed out ram's horn, was blown. This is the same instrument that was blown every 50 years on the Day of Atonement to signify the beginning of a Jubilee year. Since the shophar was blown every year on the day known as the Feast of Trumpets, we see that this feast is associated with the Jubilee.

The Jubilee year was a very special year in ancient Israel. It was a year of liberty for people and land. If a man had become poor and had sold himself to be a servant, he was to be set at liberty in the Jubilee, as we read in Leviticus 25:39-42:

> **And if thy brother *that dwelleth* by thee be waxen poor, and be sold unto thee; thou shalt not compel him to serve as a bondservant: *But* as an hired servant, *and* as a sojourner, he shall be with thee, *and* shall serve thee unto the year of jubile: And *then* shall he depart from thee, *both* he and his children with him, and shall return unto his own family, and unto the possession of his fathers shall he return. For they *are* my servants, which I brought forth out of the land of Egypt: they shall not be sold as bondmen.**

Leviticus 25:10 also tells us what happened in the Jubilee year:

> **And ye shall hallow the fiftieth year, and proclaim liberty throughout *all* the land unto all the inhabitants thereof: it shall be a jubile unto *you;* and ye shall return every man unto his possession, and ye shall return every man unto his family.**

When we read the New Testament, we find that the Lord Jesus identifies His ministry and even Himself with the year of Jubilee. We see this when we read Luke 4:18-19:

> The Spirit of the Lord *is* upon me, because he hath anointed me to preach the gospel to the poor; he hath sent me to heal the brokenhearted, to preach deliverance to the captives, and recovering of sight to the blind, to set at liberty them that are bruised, To preach the acceptable year of the Lord.

The Lord Jesus began His public ministry when John the Baptist announced Him as the Lamb of God. We read about this in John 1:29:

> The next day John seeth Jesus coming unto him, and saith, Behold the Lamb of God, which taketh away the sin of the world.

Amazingly, as we have seen, it was on the feast day known as the Feast of Trumpets in the year 29 A.D. that the Lord Jesus was announced; and so we have fulfillment of that feast day in the New Testament.

When God was about to visit the last of the plagues upon the Egyptians, He commanded the children of Israel to keep the feast known as Passover. This they did by killing a young male lamb in the evening and sprinkling its blood on the side posts and the upper door post of their house. They were also commanded to eat the flesh of the lamb that night. In the New Testament, we discover that the Lord Jesus was crucified in the year 33 A.D. on the day of Passover. We know that it was a Passover by reading, for example, John 18:39, in which Pilate addresses the Jews who wanted to crucify the Lord Jesus:

> But ye have a custom, that I should release unto you one at the passover: will ye therefore that I release unto you the King of the Jews?

As we continue to read the account of the Lord's suffering in this book of the Bible, we learn that right after he asked this question, Pilate had the Lord scourged, and then handed Him over to be crucified. The Passover lamb that the children of Israel were commanded to sacrifice was a picture of the Lord Himself. Just as the lamb's blood on their door posts protected them against the plague that killed all the firstborn of Egypt, God's elect are protected against His wrath by the blood - that is, the sacrifice - of the Lord Jesus.

The Passover was followed by seven days known as the Days of Unleavened Bread. We read about these days in Exodus 12:15-16:

> **Seven days shall ye eat unleavened bread; even the first day ye shall put away leaven out of your houses: for whosoever eateth leavened bread from the first day until the seventh day, that soul shall be cut off from Israel. And in the first day** *there shall be* **an holy convocation, and in the seventh day there shall be an holy convocation to you; no manner of work shall be done in them, save** *that* **which every man must eat, that only may be done of you.**

What are we to make of this as it pertains to God's salvation plan? In the New Testament, we find the Lord Jesus warning His disciples against the leaven of the Pharisees, as we read in Mark 8:15:

> **And he charged them, saying, Take heed, beware of the leaven of the Pharisees, and** *of* **the leaven of Herod.**

The Pharisees were religious leaders of the Jews, and Herod was their king. The Pharisees had the reputation of being meticulous about keeping God's law, so what could the Lord mean by warning His disciples to beware of the leaven of the Pharisees? The disciples wondered about this too, but then they understood, as we read in Matthew 16:12:

> **Then understood they how that he bade** *them* **not beware of the leaven of bread, but of the doctrine of the Pharisees and of the Sadducees.**

Their doctrine, or teaching, of rigorously keeping all the Old Testament laws led them to be self-righteous. In Matthew 23:23, we read what the Lord had to say about these men:

> **Woe unto you, scribes and Pharisees, hypocrites! for ye**

pay tithe of mint and anise and cummin, and have omitted the weightier *matters* of the law, judgment, mercy, and faith: these ought ye to have done, and not to leave the other undone.

We see an example of this self-righteousness in the parable the Lord told about the Pharisee and the tax collector in Luke 18:9-14:

And he spake this parable unto certain which trusted in themselves that they were righteous, and despised others: Two men went up into the temple to pray; the one a Pharisee, and the other a publican. The Pharisee stood and prayed thus with himself, God, I thank thee, that I am not as other men *are*, **extortioners, unjust, adulterers, or even as this publican. I fast twice in the week, I give tithes of all that I possess. And the publican, standing afar off, would not lift up so much as** *his* **eyes unto heaven, but smote upon his breast, saying, God be merciful to me a sinner. I tell you, this man went down to his house justified** *rather* **than the other: for every one that exalteth himself shall be abased; and he that humbleth himself shall be exalted.**

Just as leaven "puffs up" bread, we are warned not to be puffed up ourselves; but this isn't just a warning against the kind of pride or self righteousness that makes a man think himself better or holier than others. The Days of Unleavened Bread came right after the Passover and are connected with the Passover. When the Lord died - on Passover - He did all the work needed to save any one. The Days of Unleavened Bread give us a warning not to have the kind of self righteousness that makes us think we can do anything to save ourselves. It is a warning against the false gospel that is all over the churches today! It is a warning not to think that we can make even the smallest contribution toward our own salvation.

Notice that the command regarding the Days of Unleavened Bread in Exodus chapter 12 is not just to "put away leaven out of your houses," but to "eat unleavened bread" for seven days. In the New Testament, we learn that the Lord Jesus is the Bread of Life, as we read in John 6:48-51:

> I am that bread of life. Your fathers did eat manna in the wilderness, and are dead. This is the bread which cometh down from heaven, that a man may eat thereof and not die. I am the living bread which came down from heaven: if any man eat of this bread, he shall live for ever: and the bread that I will give is my flesh, which I will give for the life of the world.

Further on in this chapter, in John 6:54, we read:

> Whoso eateth my flesh, and drinketh my blood, hath eternal life; and I will raise him up at the last day.

The Lord won't be able to raise anyone up on the last day unless He's alive to do it. We know He's alive because we know He rose from the dead that Sunday, following His burial on Friday. The Days of Unleavened Bread point to His resurrection! Just as the Days of Unleavened Bread immediately followed the Passover and were linked to it, we have the resurrection immediately following the Crucifixion in God's salvation plan. In this way also, we see fulfillment of the Days of Unleavened Bread in a New Testament event.

The Feast of Pentecost - as it was known in the New Testament - was known to the ancient Israelites as the Feast of Firstfruits, as the Feast of Harvest, and as the Feast of Weeks. The date of this feast was determined by counting 50 (as signified by the prefix "pente") days after the Passover. The 50 days could also be considered as seven weeks plus one day, as we read in Leviticus 23:15:

> And ye shall count unto you from the morrow after the sabbath, from the day that ye brought the sheaf of the wave offering; seven sabbaths shall be complete:

We have already learned that it was on Pentecost that the Holy Spirit was poured out and the church age began. Importantly, this harvest of first fruits - which is associated with the church age - was small compared with the harvest that came later in the year. We see from this feast day that the "harvest" of those who are to be saved during the church age is small by comparison with the number that are to be saved toward the end of God's salvation plan.

The pouring out of the Holy Spirit on Pentecost in 33 A.D. was made possible by the Lord Jesus, as we read in John 16:7:

> **Nevertheless I tell you the truth; It is expedient for you that I go away: for if I go not away, the Comforter will not come unto you; but if I depart, I will send him unto you.**

So we see the Lord Jesus involved in this feast also.

The Feast of Tabernacles was the last annual feast to occur in the year according to the calendar followed by ancient Israel. It is the only feast for which we do not see a fulfillment in the New Testament. God's salvation plan, however, won't be over until the end of the world. By then, the events pictured by this feast will also have been fulfilled.

The Feast of Tabernacles is also called the Feast of Booths. We read something about this interesting feast in Leviticus 23:42-43:

> **Ye shall dwell in booths seven days; all that are Israelites born shall dwell in booths: That your generations may know that I made the children of Israel to dwell in booths, when I brought them out of the land of Egypt: I *am* the LORD your God.**

God is here calling our attention to the time when He led the children of Israel through the wilderness by a pillar of cloud and a pillar of fire, as we read in Exodus 13:21-22:

> **And the LORD went before them by day in a pillar of a cloud, to lead them the way; and by night in a pillar of fire, to give them light; to go by day and night: He took not away the pillar of the cloud by day, nor the pillar of fire by night, *from* before the people.**

Elsewhere in the Old Testament, in Numbers 9:21-23, we also read about the Lord leading the children of Israel in this way:

> **And *so* it was, when the cloud abode from even unto the**

> morning, and *that* the cloud was taken up in the morning, then they journeyed: whether *it was* by day or by night that the cloud was taken up, they journeyed. Or *whether it were* two days, or a month, or a year, that the cloud tarried upon the tabernacle, remaining thereon, the children of Israel abode in their tents, and journeyed not: but when it was taken up, they journeyed. At the commandment of the LORD they rested in the tents, and at the commandment of the LORD they journeyed: they kept the charge of the LORD, at the commandment of the LORD by the hand of Moses.

Here we see that God is using a cloud - and He is creating and controlling it - to command the children of Israel to either stay put or to move on. Notice that God's commandment is being associated with the cloud; but God's commandment is God's law.

This cloud must have provided protection and great relief from the scorching sun during all the time the Israelites were in the wilderness, as we read in Psalm 105:39:

> He spread a cloud for a covering; and fire to give light in the night.

This miraculous phenomenon also appears to have saved the Israelites from attack by Pharaoh's army before they could cross the sea, as we read in Exodus 14:19-20:

> And the angel of God, which went before the camp of Israel, removed and went behind them; and the pillar of the cloud went from before their face, and stood behind them: And it came between the camp of the Egyptians and the camp of Israel; and it was a cloud and darkness *to them*, but it gave light by night *to these*: so that the one came not near the other all the night.

We have seen that the cloud covered and protected the Israelites. In Isaiah 4:5-6, we see the concept of the tabernacle being associated with the cloud by day and the fire by night:

> And the LORD will create upon every dwelling place of mount Zion, and upon her assemblies, a cloud and smoke by day, and the shining of a flaming fire by night: for upon all the glory *shall be* a defence. And there shall be a tabernacle for a shadow in the daytime from the heat, and for a place of refuge, and for a covert from storm and from rain.

We therefore have an association between the tabernacle and this miraculous phenomenon, and another association between the phenomenon and God's law. Can we say that God is leading us to reason that the tabernacle represents God's law (the Bible)? We can! We may say that the Feast of Tabernacles is the feast of the Bible.

We see confirmation that the Feast of Tabernacles pertains to the Bible when we read Daniel 12:8-10:

> And I heard, but I understood not: then said I, O my Lord, what *shall be* the end of these *things*? And he said, Go thy way, Daniel: for the words *are* closed up and sealed till the time of the end. Many shall be purified, and made white, and tried; but the wicked shall do wickedly: and none of the wicked shall understand; but the wise shall understand.

When we realize that this feast pertains to the end of God's salvation plan, and that it is then that God opens up Biblical truth to a fuller understanding, we see that we may indeed view the Feast of Tabernacles as the feast of the Bible. The Bible, however, is the word of God - and the Lord Jesus is identified as the Word of God. Therefore, as with the other feasts, we see that the Feast of Tabernacles points to the Lord Jesus.

The Feast of Tabernacles also has to do with the harvest, as we read in Deuteronomy 16:13-15:

> Thou shalt observe the feast of tabernacles seven days, after that thou hast gathered in thy corn and thy wine: And thou shalt rejoice in thy feast, thou, and thy son, and thy daughter, and thy manservant, and thy

maidservant, and the Levite, the stranger, and the fatherless, and the widow, that *are* within thy gates. Seven days shalt thou keep a solemn feast unto the LORD thy God in the place which the LORD shall choose: because the LORD thy God shall bless thee in all thine increase, and in all the works of thine hands, therefore thou shalt surely rejoice.

The harvest at this time of the year is the main harvest - not the smaller harvest (the firstfruits) that is associated with the Feast of Pentecost. In Leviticus 23, after commanding Moses to keep the feasts of the Lord - including the last one listed, being the feast of tabernacles - we read the following in Leviticus 23:39:

Also in the fifteenth day of the seventh month, when ye have gathered in the fruit of the land, ye shall keep a feast unto the LORD seven days: on the first day *shall be* a sabbath, and on the eighth day *shall be* a sabbath.

Based on this scripture, we can understand how these days have also come to be known as the feast of ingathering and may be considered a feast to be held simultaneously with the Feast of Tabernacles. The Lord Jesus, by His atoning sacrifice, has made it possible for a great multitude - as pictured by the harvest brought in at the time of the feast of ingathering - to be saved.

The Feast of Tabernacles gives us a picture of what will happen at the end of God's salvation plan. There will be a time of great tribulation as God brings His judgment on the world. Everyone who is not saved by then will be subject to God's wrath, because the word of God (the Bible, which is the Law of God) condemns them. Those whom He has saved - and there will be a great harvest of them - will be protected, as promised by God's word. We read something about this in Psalm 27:5:

For in the time of trouble he shall hide me in his pavilion: in the secret of his tabernacle shall he hide me; he shall set me up upon a rock.

Elsewhere in the Bible, the Lord Jesus is referred to as the rock, and so we should think of Him when we read this scripture.

Incredibly, the entire focus of these annual feasts is on the Lord Jesus. It is so sad that, even today, many Jews all around the world are so careful to keep these feast days but are unaware of their significance. They just don't know that all of these days have to do with the saving work of the Lord Jesus. Even in the days of the prophet Isaiah, God revealed that He would not permit the Jews to understand. In Isaiah 6:9-12, we read something the Lord said to Isaiah about this:

> **And he said, Go, and tell this people, Hear ye indeed, but understand not; and see ye indeed, but perceive not. Make the heart of this people fat, and make their ears heavy, and shut their eyes; lest they see with their eyes, and hear with their ears, and understand with their heart, and convert, and be healed. Then said I, Lord, how long? And he answered, Until the cities be wasted without inhabitant, and the houses without man, and the land be utterly desolate, And the LORD have removed men far away, and *there be* a great forsaking in the midst of the land.**

Those words were a message God gave to Isaiah to deliver to ancient Judah; but the message was also a prophecy applying to all the Jews who have lived since then. They will not come to understanding, except for those relatively few on whom God has mercy. This will be their situation right up until the end of the world when there is no further possibility for God's mercy, when "the land be utterly desolate."

This message from God also seems to apply to the majority of today's Christians - and there are about two billion of them. In Isaiah 6:13, we read about God's plan to save a small number out of the whole:

> **But yet in it *shall be* a tenth, and *it* shall return, and shall be eaten: as a teil tree, and as an oak, whose substance *is* in them, when they cast *their leaves: so* the holy seed *shall be* the substance thereof.**

The tenth part identified in this scripture may apply to the vast number of people who claim to have some relationship with the Lord Jesus Christ today. The tenth part also seems to identify with the word "remnant" as used in some scriptures. For example, in Zephaniah 2:9, we read:

> **Therefore *as* I live, saith the LORD of hosts, the God of Israel, Surely Moab shall be as Sodom, and the children of Ammon as Gomorrah, *even* the breeding of nettles, and saltpits, and a perpetual desolation: the residue of my people shall spoil them, and the remnant of my people shall possess them.**

What we have learned about the annual feasts days is consistent with the conclusion that the church age comes to an end. The Feast of Pentecost, when the church age began, was called the Feast of Firstfruits by ancient Israel. The first fruits harvest, however, was a small one. The Feast of Pentecost thus points to the relatively small harvest of souls God had planned for the church age. The harvest at the end of the year, according to ancient Israel's calendar (pictured by the Feast of Tabernacles), was a big harvest. This points to the great harvest of souls God has planned following the end of the church age, toward the end of His salvation plan. (This great harvest also coincides with the time of the latter rain we read about in scripture.) We read about the great harvest in Revelation 7:9:

> **After this I beheld, and, lo, a great multitude, which no man could number, of all nations, and kindreds, and people, and tongues, stood before the throne, and before the Lamb, clothed with white robes, and palms in their hands;**

This great multitude is in white, signifying that they have been saved. In his vision, the apostle John is asked about this great multitude. In Revelation 7:14, we read:

> **And I said unto him, Sir, thou knowest. And he said to me, These are they which came out of great tribulation, and have washed their robes, and made them white in the blood of the Lamb.**

John didn't know who the great multitude could be, but he was told that they came out of great tribulation. This is the great tribulation spoken of by the Lord Jesus in Matthew 24:21:

For then shall be great tribulation, such as was not since the beginning of the world to this time, no, nor ever shall be.

This is the great tribulation that occurs toward the end of God's salvation plan, when Satan is ruling in the church; and when Satan is ruling in the church, the church age has ended!

We have seen that, according to God's salvation plan, there would come a time when true believers are to leave the church. We know that this will happen when Satan is ruling in the church, and that it will be near the end of God's plan. Can we possibly be more specific about this time? Amazingly, we can be very specific about it - as we shall see.

Chapter 8

God's Calendar For Our Day

We have previously seen that God gave us a calendar when He wrote the Bible. By using this calendar and the rules God gives us regarding numbers that appear in the Bible, we find - much to our amazement - that the calendar continues into our day. This "extended calendar" is very specific about times for the final events in God's salvation plan - events that are in the very near future.

Time Confirmation in a Parable of the Fig Tree

We need to remember that the Lord Jesus spoke in parables. Only if we keep this in mind can we understand much of what He said. We get further confirmation that we are close to the end of time by examining some parables of the fig tree. The fig tree represents national Israel. In Luke 13:6-9, we read one of these parables:

> **He spake also this parable; A certain *man* had a fig tree planted in his vineyard; and he came and sought fruit thereon, and found none. Then said he unto the dresser of his vineyard, Behold, these three years I come seeking fruit on this fig tree, and find none: cut it down; why cumbereth it the ground? And he answering said unto him, Lord, let it alone this year also, till I shall dig about it, and dung *it*: And if it bear fruit, *well*: and if not, *then* after that thou shalt cut it down.**

Notice that in the parable, the owner of the vineyard looked for fruit on the fig tree for three years. Interestingly, the length of the Lord's ministry was about three and a half years. The Lord never did find fruit in the Jewish nation, and so in another parable the Lord Jesus shows that the time would come when God would finish using them to represent the kingdom of God. In Matthew 21:19 we read:

> **And when he saw a fig tree in the way, he came to it, and found nothing thereon, but leaves only, and said unto it, Let no fruit grow on thee henceforth for ever. And presently the fig tree withered away.**

We know that not long after this incident, God began using the churches to send out the Gospel. We also know from history that the Jewish nation ceased to exist after 70 A.D. because it was destroyed by the Romans. There was no more fig tree! Yet the Lord Jesus told another parable to give us an indication of when the end times arrive. The Lord had been speaking to His disciples about the last days, as we read in Mark 13:27:

> **And then shall he send his angels, and shall gather together his elect from the four winds, from the uttermost part of the earth to the uttermost part of heaven.**

This discourse continues in the next two verses, in Mark 13:28-29, where we read this remarkable parable:

> **Now learn a parable of the fig tree; When her branch is yet tender, and putteth forth leaves, ye know that summer is near: So ye in like manner, when ye shall see these things come to pass, know that it is nigh, *even* at the doors.**

What is this parable telling us? It is now understood by many to mean that as we get close to the end of time, national Israel would once again exist. However, the fig tree - Israel - would only be "in leaf." There would be no fruit, spiritually speaking. In fact, in the year 1948 and against all odds, the nation of Israel did once again come into existence. Therefore this parable indicates that we are

1994 is a Key Year

We also know that, before the end comes, God will save a great multitude which no man could number. We might suspect that this surge in salvation activity would occur in a Jubilee year. Earlier, we learned that a Jubilee year occurred every 50 years. We learned that the year 7 B.C., when the Lord Jesus was born, was a Jubilee year. If we count 50 year intervals from 7 B.C., we see that the next Jubilee year occurred in 44 A.D., then in 94 A.D., and so on. Therefore, the year 1944 was a Jubilee year, and so was the year 1994. The first Jubilee year to occur after 1948 - the year that the "fig tree" was once again in leaf - was 1994. That year, 1994, was a key year in God's salvation plan. But what exactly happened then? There was no "earth shattering" event that anyone noticed.

As a matter of fact, the Bible indicates that something very important did happen then. It was something that was happening spiritually, so there was no visible evidence.

There are many scriptures dealing with the final days, and it is no easy matter to sort them all out. On the one hand, we learn that the great tribulation will be in those days. We also know that very many people will be saved then, and that God's wrath will come on the unsaved. In Matthew 24, where we read the Lord's command to "flee to the mountains," we also read a reference to the prophet Daniel. There, the Lord also speaks of the "abomination of desolation" in the holy place.

The Book of Daniel has a great deal to tell us about end time events, and in Daniel 8, we read something that should suggest to us the Lord's words from Matthew 24. In Daniel 8:13-14, we read:

> **Then I heard one saint speaking, and another saint said unto that certain *saint* which spake, How long *shall be* the vision *concerning* the daily *sacrifice*, and the transgression of desolation, to give both the sanctuary**

and the host to be trodden underfoot? And he said unto me, Unto two thousand and three hundred days; then shall the sanctuary be cleansed.

According to this scripture, a period of 2,300 days is somehow involved in end time events. We have focused our attention on 1994 as being a key year occurring near the end of God's salvation plan. It doesn't seem likely that the 2,300 day period - a time when the sanctuary is "trodden underfoot" - would begin in a Jubilee year, which 1994 was. Suppose, however, that the 2,300 days ended then. If we divide 2,300 by the number of days in a year, we see that it's a little more that six years and three months. Therefore we might expect that the 2,300 days of this prophecy began in 1988 or so.

In considering 1994 and its relation to 1877 B.C. and 587 B.C., we see that it is to be expected that the tribulation of our time began earlier than 1994. After all, Jacob's tribulation - a result of famine across the land - began well before he left his homeland. Similarly, the nation of Judah was seriously weakened years before they were conquered. In this way, the three tribulations are consistent and fit the same pattern if we understand that 1994 ended the 2,300 day period.

Another Trail Leads to 1994

The Bible brings our attention to 1994 in another way. We know that the Bible speaks of great tribulation in connection with end time events. Also, we learned that in the year 1877 B.C., Jacob went to Egypt during his time of great tribulation; and we learned that the nation of Judah was destroyed in 587 B.C. - a time of God's great wrath on His people. The time interval between these two events is 1,290 years. By following a trail using several Biblical relationships, we will again arrive at 1994.

First, we find that the number 1,290 appears in the Bible in Daniel 12:11:

And from the time *that* the daily *sacrifice* shall be taken

away, and the abomination that maketh desolate set up, *there shall be* a thousand two hundred and ninety days.

This scripture speaks of 1,290 days, not years; but God has given us a precedent of using a year for a day. In the Book of Numbers, the spies searched out the land of Canaan for forty days. Instead of acting to take the land as they were supposed to, the people were fearful; and so God kept them in the wilderness for forty years - a year of wandering for each day of searching out the land. Now, notice in this scripture the reference to the "abomination that maketh desolate." This suggests the Lord's discourse in Matthew 24, in which He spoke of great tribulation in the last days. Between Matthew 24 and Daniel 12:11, we have an association between the number 1,290 and the great tribulation of the last days. And using the year-for-a-day principle, we understand the 1,290 days to represent 1,290 years.

There is another Biblical relationship of which we need to be aware. We may call it the "one part, two parts" relationship. We see this relationship applied to people in Zechariah 13:8:

> **And it shall come to pass, *that* in all the land, saith the LORD, two parts therein shall be cut off *and* die; but the third part shall be left therein.**

We can also see this relationship applied to time in the following way. The time interval between the Exodus (1447 B.C.) and the Lord's birth (7 B.C.) is 1,440 years. Comparing this 1,440 years with the time interval between the Exodus and the year that Solomon began to build the temple (480 years, according to 1 Kings 6), we see that the time from the Exodus to the start of the earthly temple was one part or a third, and the time from the start of the earthly temple to the Lord's birth was two parts or two thirds. We should remember what the Lord Jesus told the Jews in John 2:19:

> **Jesus answered and said unto them, Destroy this temple, and in three days I will raise it up.**

In this scripture, the Lord was comparing Himself with the earthly temple of His day. So we see this "one part, two parts" relationship applied to time (480 years and 960 years as fractions of 1,440 years).

Let us now go back to the 1,290 year time interval between Jacob's tribulation in 1877 B.C., and the destruction of the temple in 587 B.C. If we take the interval of 1,290 years to be the "one part," then two parts would be 2 x 1,290 years, or 2,580 years. Advancing the calendar from 587 B.C. by 2,580 years brings us to 1994, giving us confirmation of that date.

What Happened Before 1994?

We have seen that 1994 was the end of the 2,300 days of Daniel 8. Let's assume that the Bible is directing us to a particular day in the year 1988 as the beginning of this period. If this assumption is correct, we should find confirmation of it in the Bible. What day of the year 1988 could it possibly be? The day we're looking for is the day the church age ended. It is the day when the abomination of desolation came into the holy place - that is, Satan began to rule in the church.

We know that the church age began on Pentecost in 33 A.D., so we might suspect that the Bible is directing us to the day before Pentecost. In 33 A.D., Pentecost fell on May 22. Interestingly, in 1988 it also fell on May 22. If the church age ended the day before Pentecost in 1988, then it ended on May 21. This would be the beginning of the 2,300 day period that is identified with great tribulation. Let us test this assumption for Biblical confirmation. If we count 2,300 days from May 21, 1988, we arrive at September 7, 1994. Is there anything special about *that* day? Indeed, there is! It turns out that in 1994, September 7 was the date of the feast known as the Feast of Trumpets. This is the feast associated with the Jubilee. Therefore, the 2,300 days fit exactly into the time interval between May 21, 1988 and September 7, 1994; and this time interval is bracketed by two dates that are identified with feast days in the ceremonial calendar. This certainly confirms that the church age ended on May 21, 1988. Also, it is significant that this time interval - from the beginning of the church age in 33 A.D. to its end in 1988 - is exactly 1,955 years long. This number breaks down into the following numbers of spiritual significance:

5 x 17 x 23 (duration of church age)

For 1,955 years, God used local congregations as the vehicle to distribute His word (the Bible) around the world. This period began on a wonderful day in the year 33 A.D., when God saved about three thousand people. It ended on a sad day in 1988, without even being noticed, when God gave up on using the church and allowed Satan to rule there. Earlier, we saw that the number 5 identifies with the atonement, which emphasizes both judgment and salvation; the number 17 emphasizes heaven, and the number 23 emphasizes God's wrath or judgment. Therefore - by these factors for 1,955 - God may be telling us that He has applied the atonement to save those whom He would by using the church, but that He ultimately brought His judgment on it.

The Great Tribulation Continues

If we were working through the Bible as we are doing at any time prior to 1994, we might conclude that 1994 would be the end of the world. After all, we discovered that it was a key year in God's salvation plan. We know that God promised to save a great multitude which no man could number, but He could do that very quickly if He chose to. From what we read in the Bible, we may think that He could have done it in a single day - even in 1994 on that September 7 that ended the 2,300 day period after the end of the church age. Obviously, that's not what God planned.

Since time has continued beyond 1994, we know that God's plan continues beyond then. In order for us to see if God may have revealed any dates beyond 1994, we need to again examine the two tribulations we considered earlier.

The year 1877 B.C. was for Jacob actually a milestone in a seven year tribulation. The famine that caused him to leave his homeland had begun two years earlier. In Genesis 45, Joseph -

who had been sold into slavery by his brothers years earlier but had risen to great power in Egypt - revealed his identity to his brothers and told them that the famine would continue for five more years. In Genesis 45:6, we read:

> **For these two years** *hath* **the famine** *been* **in the land: and yet** *there are* **five years, in the which** *there shall* **neither** *be* **earing nor harvest.**

God had made these things known to Joseph, and caused him to be in a position to make provisions against the famine. We know that Jacob came down quickly to Egypt after Joseph spoke to his brothers on that occasion. Therefore, Jacob's departure came two years into a seven year tribulation, or 2/7 of the way to the end of that tribulation.

Now, let's take another look at what happened in 587 B.C. and the years shortly before and after. Jeremiah had prophesied a 70 year period of "desolations of Jerusalem," to use the words of Daniel 9. The fall of Jerusalem in 587 B.C., as horrendous as it was, did not actually end this tribulation as far as the Bible is concerned. We learned that this period began in the year 609 B.C. when the Egyptian pharaoh removed Jehoahaz as king of Judah; it continued until 539 B.C. - the year Babylon fell. Therefore, the fall of Jerusalem occurred 22 years after this tribulation had begun, or 22/70 of the way to the end of it. This ratio is reasonably close to the ratio we discovered when examining Jacob's tribulation. If the pattern holds for the great tribulation of our own time, it means that there is a number such that the ratio of 2,300 to it is reasonably close to 2/7 and to 22/70. Each of these ratios is about 30%, so we would expect our ratio to also be about that much. Also, the number we're looking for may be a multiple of seven as are the other two denominators. At a guess, we could try the number 7,000 (7 x 10 x 10). However, as it turns out, the number 8,400 (which is 7 x 12 x 10 x 10) brings us to a date which the Bible confirms in many ways.

Let's see how this works. We learned that the 2,300 day portion of the end time great tribulation ends on September 7, 1994. These 2,300 days are the first part of 8,400 days. Therefore, there are 6,100 days remaining after September 7, 1994. When we count 6,100 days from September 7, 1994, we come to May 21, 2011. This is the day that ends the great tribulation.

We have now generated from the Bible two dates from the recent past (May 21, 1988 and September 7, 1994) and one date into the very near future (May 21, 2011). To understand the significance of these dates, we need to examine some of the scriptures dealing with end time events. There are many such scriptures. It is not always apparent, however, that a scripture deals with the end times. For example, in Matthew 24, the Lord was responding to His disciples when they had pointed out to Him the buildings of the temple complex. It must have been a very impressive sight to them. In Matthew 24:2, we read this statement by the Lord:

And Jesus said unto them, See ye not all these things? verily I say unto you, There shall not be left here one stone upon another, that shall not be thrown down.

This is one of those key scriptures that does in fact deal with end time events. The Lord's statement here has traditionally been understood to be a prophecy dealing with the physical destruction of the temple. (In fact, this did happen in 70 A.D., as we learned earlier.) Suppose, however, that there is an altogether different meaning in His statement - a meaning associated with the end of the world. We know that in many scriptures, the body of believers is compared to a physical structure. In Matthew 24:2, God appears to be doing something similar: equating the physical temple (which was identified with God's kingdom) with the future external representation of God's kingdom - the local congregations of the church age. The church age had not even begun, and already the Lord was talking about its end. Neither do we read anywhere that this structure is reassembled. The stones are thrown down by God. Therefore, true believers are to leave the church and not look back. With this in mind, we can be certain that September 7, 1994 - which ended the first part of the great tribulation - did not give an "all clear" sign for true believers to return to their local congregations.

In order for us to attempt to understand what God did in 1994 and what He has planned for the near future, we need to consider yet other scriptures dealing with the end of the world. One of these scriptures is Revelation 8:1, where we read:

> **And when he had opened the seventh seal, there was silence in heaven about the space of half an hour.**

The silence in heaven indicates that no one, or perhaps only very few persons, were being saved during the half hour. For that period of time, there was no rejoicing in heaven as there is whenever someone is saved; there was only silence. This period seems to identify with the time of darkness mentioned in Isaiah 60:2:

> **For, behold, the darkness shall cover the earth, and gross darkness the people: but the LORD shall arise upon thee, and his glory shall be seen upon thee.**

After the darkness cited in this scripture has passed from the earth, God resumes His program of saving people so that there is no longer silence in heaven. But are we to understand the half an hour of Revelation 8:1 as a literal period of time? We know that many times, God uses a number to represent or symbolize another number. Depending on the context and by comparing scripture with scripture, we may be able to determine where this is done. It is clear, for example, that when God told Noah to build the ark to be a certain size as recorded in Genesis, the numbers stated there are the actual dimensions of the ark. In Revelation 8:1, however the half hour appears to represent the first part of the great tribulation. We have seen that this period of time is 2,300 days. Also - as we have seen - this 2,300 day period began on May 21, 1988 with the end of the church age and continued until September 7, 1994. As soon as this period ended, God again began to save people. In fact, from what we see in the Bible, God has resumed His work and is planning to save more people than ever before in history. In Revelation 7:9, we read of the great multitude that is to be saved:

> **After this I beheld, and, lo, a great multitude, which no man could number, of all nations, and kindreds, and**

people, and tongues, stood before the throne, and before the Lamb, clothed with white robes, and palms in their hands;

In Revelation 7:14, we read something else about this great multitude:

And I said unto him, Sir, thou knowest. And he said to me, These are they which came out of great tribulation, and have washed their robes, and made them white in the blood of the Lamb.

Notice that the great multitude came out of great tribulation. Notice also that, according to Revelation 7:9, it is a great multitude which no man could number. According to the almanac, the earth's population is somewhat over six and a half billion people at the time of this writing. No one knows the exact number of course, but nevertheless a number has been applied to the earth's population - just as a number has been applied to the number of persons identifying themselves as Christians (approximately two billion persons). Could it be that God used the words "which no man could number" because the great multitude does not belong to the world's Christian congregations where they may be counted? In fact they may be scattered throughout all nations, including those that are hostile to Christianity. The Bible gives us indications that this is the case.

The Final Feast of Tabernacles

We have seen that the Bible points to May 21, 2011 as the day that ends the great tribulation. We have also seen that God has used the annual feast days as turning points in His salvation plan. Yet May 21 of 2011 is not one of these feast days. We also know that of all the annual feast days given to ancient Israel, each one has had its fulfillment in the New Testament - except the Feast of Tabernacles. Therefore, we should expect that the Feast of Tabernacles will indeed be fulfilled in some way before the end of the world. When we read John 7:37, we get a clue that the fulfillment may be that the world will end on that date:

> In the last day, that great *day* of the feast, Jesus stood and cried, saying, If any man thirst, let him come unto me, and drink.

From John 7:2, we know that the Lord spoke those words on the occasion of the Feast of Tabernacles:

> Now the Jews' feast of tabernacles was at hand.

The Lord's words "the last day" used in connection with the Feast of Tabernacles should prompt us to want to know the dates associated with this feast. We were led to the year 2011 when we discovered the date May 21, 2011, and so this is the year in which we expect the world to end.

In the year 2011, this feast will begin on October 13 - which is the fifteenth day of the seventh month (Tishrei) of the Jewish ceremonial calendar. The eighth day of the feast, which is the day of a holy convocation or solemn assembly, will be October 20. This eighth day is a sabbath and is considered to be the last day of the feast. However, there is something very interesting we read in 1 Kings 8. This is an account of the dedication of the temple in Solomon's time. It happened during the Feast of Tabernacles, so this was a very special Feast of Tabernacles. In 1 Kings 8:66, we read:

> On the eighth day he sent the people away: and they blessed the king, and went unto their tents joyful and glad of heart for all the goodness that the LORD had done for David his servant, and for Israel his people.

This scripture is surprising because the eighth day - which was the twenty-second of the month in the Jewish ceremonial calendar - was a Sabbath day, and the people were not permitted to travel the kinds of distances they would have had to travel to arrive at Jerusalem. After all, the people had come from all over the kingdom. In sending the people away, Solomon would be sending them on their way to return home. So what are we to make of this? There is another account of the temple dedication in 2 Chronicles that we also need to consider. In 2 Chronicles 7:10, we read:

> And on the three and twentieth day of the seventh month he sent the people away into their tents, glad and merry in heart for the goodness that the LORD had shewed unto David, and to Solomon, and to Israel his people.

In this account, Solomon sent the people away on the twenty-third day. When we compare the two accounts, we see that God is linking the eighth day to the twenty-third day of the month so that the eighth day extends to the following day. When we take this clue into account, we come to the tentative conclusion (we will see proof that it is a correct conclusion) that the last day of the Feast of Tabernacles - which God is designating to be October 21 in 2011 - will be the last day of the earth's existence.

A Review of the Path to Discovering the Dates

Using October 21, 2011 as the date for the end of the world, we may now go back to the earlier dates we discovered - examining scriptures dealing with end times and looking for clues that may help us to understand the significance of these earlier dates. Before proceeding, however, let's briefly summarize the dates we discovered and review the path we followed to arrive at them.

We know that the great tribulation began when the church age ended on May 21, 1988, and that it is in two parts: the first part ended on September 7, 1994 and the second part will end on May 21, 2011. We also know that the world will end on October 21, 2011. Here is how we arrived at these dates:

1. We began on the path to discover these dates knowing that there would come a time, toward the end of time, when the church age ended.

2. We also knew that an important indicator that we are near the end of time was that "the fig tree" would "put forth

leaves." That happened when Israel was re-born in 1948 as a nation.

3. We then determined that the first Jubilee year to occur after 1948 was the Jubilee year of 1994. Therefore, we knew that 1994 was a key year in God's plan.

4. We made a connection between the Lord's discourse in Matthew 24 with a scripture from the Book of Daniel. In Matthew 24, the Lord spoke of the "abomination of desolation" and gave His command to "flee to the mountains," that is, to leave the church. In the Book of Daniel, we read of 2,300 days of "transgression of desolation," which suggests Matthew 24.

5. It did not seem likely that this terrible period of 2,300 days would begin in a Jubilee year (1994), so we suspected that it ended then. Starting in 1994 and counting back in time by 2,300 days brought us to 1988 as the likely year the "transgression of desolation" began.

6. We focused our attention on May 21, 1988 because it was the day before the Pentecost of that year. (Remember, the church age began in 33 A.D. on the day of Pentecost.)

7. Counting forward by 2,300 days from that date, we arrived at September 7, 1994 - and that turns out to be the exact date on which the feast known as the Feast of Trumpets occurred.

8. We then tried to determine the duration of the final (great) tribulation. We examined two earlier tribulations which each had two parts, and looked for a number having two characteristics: first, that it featured the number 7, just as the total duration for each of the other two tribulations does; second, that it should be a number such that the ratio of 2,300 days to that number was approximately the same as the ratio we get for each earlier tribulation by comparing its first part to its total duration. The number we arrived at was 8,400. This is the projected length in days for the current great tribulation.

Chapter 8 God's Calendar For Our Day

9. Since the first part of the current tribulation is 2,300 days and its total length is 8,400 days, there are 6,100 days in the second part. We therefore were able to determine that the second part began on September 7, 1994.

10. Counting forward by 6,100 days from September 7, 1994 brought us to May 21, 2011 as the day that ends the current great tribulation.

11. With our attention focused on 2011, and realizing that the Feast of Tabernacles is the only feast of ancient Israel that has not had its fulfillment in the New Testament, we picked the day that will be the last day of that feast in 2011 as the day that will likely be the end of the world.

12. The Feast of Tabernacles ordinarily ends on its eighth day, which will be on October 20 in 2011; but we examined an account of a very special Feast of Tabernacles - the one that took place at the dedication of Solomon's temple. In the Bible's account of that feast, the eighth day was linked to the day that followed it. The eighth day, therefore, appeared to be counted as if it were two days. This led us to conclude that October 21, 2011 will mark the end of the world.

We may be certain that these end-time dates were discovered only after much checking and re-checking, and only after a great deal of prayer. Although the path we followed to arrive at these dates may seem tenuous, we shall see that there is solid proof that the dates are correct. Before we consider this proof, however, let us try to understand what the Bible is telling us about the dates we discovered.

We learned that God had been using the local congregations to get the Gospel out to the world for 1,955 years, but that this all ended on May 21, 1988. We know that from that date until September 7, 1994, hardly anyone was being saved; but beginning on September 7, 1994, God once again began to save people. In fact, we know that in the period beginning on that day and continuing until May 21, 2011, God has promised to save a great many people. These people are being saved outside of the local congregations. Five months after May 21 in 2011 - on October 21 the end will come.

Possibilities for 1994

The year 1994, as we have seen, ends the first part of the on-going 8400 day great tribulation. By analyzing Jacob's tribulation and the tribulation which ended the nation of Judah, we saw that the year 1994 should have been a year in which something terrible happened. In each of the two earlier tribulations, there is one year that stands out in the tribulation and divides it into two parts. In Jacob's tribulation, the year was 1877 B.C. - the year Jacob was commanded by God to leave the promised land, Canaan and go to live in Egypt. That must have brought great sorrow to Jacob and his family. In Judah's tribulation, the year 587 B.C marked the year that the Babylonians destroyed Jerusalem and the magnificent temple - a horrible time for anyone who was there. In the current great tribulation, it is the year 1994 that corresponds to those other two years that stand out in their respective tribulations. What terrible thing could have happened in 1994? Outside the local congregations, something wonderful was happening. Whatever terrible thing it was that happened then must have happened to the local congregations.

It may not be possible to know with certainty what happened to the local congregations in 1994, but there are some possibilities. The very fact that the Holy Spirit was poured out into the world, but the local congregations were excluded from this blessing, is a terrible indictment against them. This was part of God's judgment process, as we read in 1 Peter 4:17:

> For the time *is come* that judgment must begin at the house of God: and if *it* first *begin* at us, what shall the end *be* of them that obey not the gospel of God?

Perhaps God's judgment against the churches began almost at the beginning of the church age, as 1 Peter 4:17 may be indicating. In any event, the entire world suffered in the process that began on May 21 in 1988 because hardly anyone was being saved;

beginning September 7 in 1994, there was once again hope - but not for anyone who is a member of a local congregation and refuses to leave it. Therefore, 1994 must mark a milestone in God's judgment against the local congregations worldwide.

We also know that there would come a time when God would actively delude men to hide the truth. Unless He guides us to truth, we won't discover it. To think that God would actually cloud someone's mind - delude him or her - to prevent that person from gaining a blessing from His word, is a scary thought indeed. Yet this is what we read in 2 Thessalonians 2:11-12:

> **And for this cause God shall send them strong delusion, that they should believe a lie: That they all might be damned who believed not the truth, but had pleasure in unrighteousness.**

Perhaps this is what began to happen in 1994 within the local congregations. In recent years, churches have become increasingly tolerant of teachings that deviate from Biblical truth. Churches that tolerate divorce and homosexuality are glaring examples of this. This trend may be attributed, at least in part, to God's "strong delusion."

What Happens After 1994?

We have seen that God often follows a previously established pattern when He does something; and we know that He will eventually destroy the earth. To help us understand what events may lead up to this, we should consider the Biblical account of the first occasion of worldwide destruction: the flood of Noah's day. All human life perished in that flood, except for the eight persons who were on the ark. When we read the scriptures dealing with the flood, we realize that God gave mankind - through Noah - a warning that the flood was coming. In Genesis 7:1-4, we read:

> **And the LORD said unto Noah, Come thou and all thy house into the ark; for thee have I seen righteous before me in this generation. Of every clean beast thou shalt**

> take to thee by sevens, the male and his female: and of beasts that *are* not clean by two, the male and his female. Of fowls also of the air by sevens, the male and the female; to keep seed alive upon the face of all the earth. For yet seven days, and I will cause it to rain upon the earth forty days and forty nights; and every living substance that I have made will I destroy from off the face of the earth.

As we have read, Noah and his family went into the ark seven days before the rain began. After Noah and the animals were safely aboard, the door was shut. In Genesis 7:16, we read:

> **And they that went in, went in male and female of all flesh, as God had commanded him: and the LORD shut him in.**

Notice that God Himself shut the door. When the rains began, people realized that Noah -whom they probably had been ridiculing for years - had been right in what he had been preaching. They may have frantically banged on the door to the ark, begging to be let in; but it was too late. The door was shut.

The rain continued for forty days and forty nights. The Bible doesn't tell us how long it took for the last person to die. We may safely assume that at least some of those who were on higher ground managed to survive longer than those who could not escape the low lying areas. Eventually, however, all of mankind - except for Noah and his family - perished. In Genesis 7:21-22, we read:

> **And all flesh died that moved upon the earth, both of fowl, and of cattle, and of beast, and of every creeping thing that creepeth upon the earth, and every man: All in whose nostrils *was* the breath of life, of all that *was* in the dry *land*, died.**

There was some interval of time between the day Noah and his family got into the ark and the day the last person died. Could there be similarities between this ancient judgment on mankind, and the way God is planning to execute sentence on the world in these end times? Indeed there is, and we will see that the account of the flood helps us to understand the significance of the last two dates we discovered.

We know that the great tribulation continues for 8,400 days, until May 21, 2011; and we know that the world will end on October 21, 2011. We also know that there are two great and unmistakable end time events yet to happen according to the Bible: the rapture and the destruction of the universe by fire. We read about the rapture in 1 Thessalonians 4:16-17:

> **For the Lord himself shall descend from heaven with a shout, with the voice of the archangel, and with the trump of God: and the dead in Christ shall rise first: Then we which are alive *and* remain shall be caught up together with them in the clouds, to meet the Lord in the air: and so shall we ever be with the Lord.**

This day of the rapture - the day when the true believers are caught up in the clouds to meet the Lord - is the day when God will "shut the door" so to speak. Until then, He will extend His mercy because this is still the day of salvation. After that, there will be no more mercy. God will begin the equivalent of what the ancients experienced when the flood waters came. There will be hell on earth until the last day. That last day of the world - October 21 in 2011 - will be the day God destroys everything by fire. We read about this in 2 Peter 3:11-12:

> ***Seeing* then *that* all these things shall be dissolved, what manner *of persons* ought ye to be in *all* holy conversation and godliness, Looking for and hasting unto the coming of the day of God, wherein the heavens being on fire shall be dissolved, and the elements shall melt with fervent heat?**

In Noah's day, God used water to destroy almost all of humanity. In our day, He will use fire to make a final end of the entire universe. In Noah's day, God used a great ark to save those whom He chose to save. In these end times, he will remove His elect from off the face of the earth and so spare them the judgment coming to the earth - a judgment that will end in complete destruction.

We may now briefly summarize what we have learned about each of the four dates we have discovered:

May 21, 1988 — End of the church age; beginning of the great tribulation.

September 7, 1994 — End of the first part of the great tribulation; this day begins the time of great harvest outside the local congregations, and is associated with judgment against the local congregations.

May 21, 2011 — The great tribulation ends; day of the rapture - true believers who are living are carried up into the clouds to meet the Lord Jesus, and true believers who have died are resurrected to join them; the unsaved dead are also resurrected, but not to life; a great earthquake begins a time of "hell on earth."

October 21, 2011 — End of the world and of the entire universe; annihilation of the unsaved.

Earlier, it was stated that there is actually proof that the dates we discovered are correct. Amazingly, the Bible verifies these dates in several dramatic ways so that there can be no doubt about them. Let us now see how the Bible does this.

Numerical Patterns Provide Numerous Proofs

We shall see that the four dates we discovered fit into the overall framework of dates previously established by the Biblical calendar. The two dates of the recent past and the two of the near future fit into this framework so perfectly that there can be no doubt that God planned it that way.

We know that the church age began in 33 A.D. and ended in 1988. Earlier, we saw that the interval between these years - 1,955 - may be broken down as follows:

5 x 17 x 23 (duration of the church age)

Chapter 8 God's Calendar For Our Day

This is one of many numerical patterns associated with the four end-time dates. As indicated earlier, this pattern seems to point to the salvation (the number 17) through the atonement (the number 5) God planned for those who would become saved throughout the church age, and the judgment (the number 23) He would bring against the local congregations when He was finished with them.

We also came across the number 2,300 in a scripture from the Book of Daniel and discovered that it was the duration (in days) for the first part of the ongoing great tribulation. It was also the time interval between the first two of our four end-time dates. This number may be broken down as:

23 x 100 (time interval in days between end of church age
and beginning of great salvation plan outside
the local congregations)

Notice the number 23, which we have seen to be associated with God's wrath or judgment. The number 100, like the number 10, is associated with the completeness of whatever is in view. God's use of the number 2,300 therefore seems to be pointing to His wrath against the local congregations.

Earlier, we came across the number 1,290 - also from the Book of Daniel - and learned that it was the time interval between key years of two past tribulations. This number is also, however, linked to the year 1994 in a "one part, two parts" relationship. Consequently, the time interval between the key year of the latter of those two past tribulations and the year 1994 is 2,580 years. This number breaks down as:

2 x 3 x 10 x 43 (time interval between 587 B.C. and 1994)

We know that the year 587 B.C. was the year Jerusalem was destroyed by the Babylonians, and this pattern appears to be linking the year 1994 to that year. By these numbers, God may be telling us that it was His complete purpose (the numbers 3 and 10) to bring His judgment (the number 43) against those whom He had commissioned (the local congregations) to bring the Gospel (the number 2), just as He brought His judgment on ancient Judah in 587 B.C. Note that the number 2,580 may also be broken down in two other ways; but these also feature the number 43, thereby indicating God's wrath or judgment.

We learned that 33 A.D. was the year of the Crucifixion and that 2011 will mark the end of the world. The difference between these two numbers is 1,978 years, which may be broken down as follows:

2 x 23 x 43 (time interval between year of Crucifixion and end of world)

Perhaps this pattern should remind us that the Lord Jesus experienced God's wrath in 33 A.D. as a demonstration of the payment made for those He came to save (the number 2); and that He will bring His judgment (the numbers 23 and 43) against everyone else at the end of the world.

When we were analyzing God's calendar, we discovered that the year 11,013 B.C. marked the Creation of the world. Notice the interval from that year to the year 2011:

13,000 + 23 (time interval between Creation and end of the world)

Again, we see the number 23, so this time interval seems to indicate that God's judgment will come after the time expanse He allotted to the world (a multiple of 1,000 years by the number 13).

Earlier, we examined Biblical evidence pointing to 7 B.C as the year of birth for the Lord Jesus. From then until the year 2011, there are 2017 years (7 + 2011 - 1). We can express this number of years as:

2,000 + 17 (time interval between the Lord's birth and the rapture)

Here we see a numerical pattern with the number 17 (indicating heaven) added to 2,000 (or 2 x 1,000) years. Recall that we have learned that the number 2 is associated with those who have been commissioned to bring the Gospel. These are certainly significant numbers for the second coming of the Lord, when He takes His

Chapter 8 God's Calendar For Our Day

elect to heaven.

We discovered that the year 391 B.C. marked the last word from God in the Old Testament. It was not until He sent John the Baptist that there was once again a message from heaven. We also learned that the Feast of Tabernacles in the year 2011 will mark the end of the world, and that this feast is associated with the Bible - so much so that it may be referred to as the feast of the Bible. By the year 2011, God will have opened His word to its fullest extent. He won't be revealing anything else to us during our lifetime. The time interval between these two dates is 2,401 years and may be factored as follows:

7 x 7 x 7 x 7 (time interval from 391 B.C. and 2011)

The complete opening up of God's word to us may be indicated by this remarkable numerical pattern, which features the number 7 (associated with the perfect fulfillment of God's purpose) and the number 4 (associated with the farthest extent in time or distance that God spiritually has in view) because of the multiples of the number 7.

When we started on the path to establishing the four dates for which we are now seeing Biblical confirmation, we learned the importance of the year 1948. The time interval between that year (when the nation of Israel was established) and the year 1994 (the first Jubilee year to follow 1948) is 46 years. This interval breaks down into:

2 x 23 (time interval from 1948 to 1994)

Perhaps this sequence is indicating God's judgment (the number 23) against those who had been entrusted to preserve the word of God (the number 2) in the past (ancient Israel, then the Jews) and against those who were more recently commissioned to bring the Gospel (the local congregations of Christian churches all around the world).

There is also a numerical pattern relating the birth of national Israel to the end of the church age. From the year modern-day Israel was established until the year the church age ended, there is an interval of 40 years:

40 years (time interval from 1948 to 1988)

What are we to make of this? Recall that Israel is identified with the fig tree. In Luke 13, the Lord spoke a parable about a fig tree. We find this parable in Luke 13:6-9:

> **He spake also this parable; A certain *man* had a fig tree planted in his vineyard; and he came and sought fruit thereon, and found none. Then said he unto the dresser of his vineyard, Behold, these three years I come seeking fruit on this fig tree, and find none: cut it down; why cumbereth it the ground? And he answering said unto him, Lord, let it alone this year also, till I shall dig about it, and dung *it*: And if it bear fruit, *well*: and if not, *then* after that thou shalt cut it down.**

Notice that the fig tree was being tested. If it didn't bear fruit after a certain time, it would be cut down. God uses the number 40 to denote testing. Based on the time interval for the dates giving us this number, we must conclude that God gave Israel until 1988 to bear spiritual fruit: to turn to the Lord Jesus Christ as their Savior. There is no evidence that this has happened, and so we must conclude that God will "cut down" national Israel, as He will everyone else who is not one of His elect.

There is still more to learn as we investigate time intervals between the date of establishment of modern Israel and dates of the final events in God's salvation plan. The exact date for the birth of Israel is May 14, 1948. If we calculate the number of days from that date until the date of the rapture (using 365.2422 days/year and dropping the fraction of a day from the product), we get 23,017 days, which equates to:

23,000 + 17 (time interval in days, May 14, 1948 to May 21, 2011)

We learned that the number 23 signifies God's wrath or judgment, the number 10 (or multiples of itself, like 1000) signifies completeness, and the number 17 signifies heaven. Therefore,

Chapter 8 God's Calendar For Our Day

these numbers seem to be telling us that there will be wrath for Israel, while God's elect are gathered from all over the earth to be taken to heaven.

Furthermore, if we calculate the number of days from the exact date of Israel's establishment until the date the Bible reveals to be the last day of the earth's existence, we get 23,170 days, which may be expressed as:

23,000 + 170 (time interval in days, May 14, 1948 to October 21, 2011)

The number 170, with its factor of 10 signifying completeness, signifies heaven just as the number 17 does. Like the preceding numerical pattern, this one indicates God's wrath - a wrath causing complete destruction - for Israel, and heaven for God's elect. In view of these patterns, it seems that God is warning Israel and all those who put their faith in the Jewish religion that they - like the local congregations of Christian churches all over the world - will be those who are "beaten with many stripes."

There is still another numerical pattern associated with the year 1948 and the final year. The time interval from 1948 until the year 2011 is 63 years. This interval breaks down as follows:

3 x 3 x 7 (time interval from 1948 to 2011)

The numbers 3 and 7 both indicate God's purpose. The rebirth of Israel in 1948 could most certainly never have happened except that it was God's purpose to bring it about. Perhaps, in these numbers, God is indicating the coming fulfillment of His purpose to save the elect by the year 2011, just as it was His purpose to resurrect the nation of Israel.

As a final pattern involving the rebirth of national Israel in 1948, consider the time interval from 1407 B.C. until that date. Recall how we learned that 1407 B.C. was the year Israel crossed the Jordan and entered the promised land. From that year until the year 1948, there is a time interval of 3,354 years, which breaks down as follows:

2 x 3 x 13 x 43 (time interval from 1407 B.C. to 1948)

This numerical pattern seems to be telling us that it would be God's purpose (the number 3) to bring His wrath or judgment (the number 43) at the end of the world (the number 13) on those who had been entrusted to preserve God's word (first the kingdom of Israel and then the Jews) or on those who have been commissioned to bring the Gospel to the world throughout the church age (the number 2).

We learned that 4990 B.C. was the year of the great flood of Noah's day. The time interval between that year and the year 2011 is 7,000 years:

7 x 1000 (time interval from the Flood until end of the world)

The number 7 indicates the perfect fulfillment of God's purpose and the number 1000 (like the number 10) indicates the completeness of whatever is in view. Appearing together in this pattern, they seem to indicate a complete chapter in the earth's history, from the first occasion of God's judgment on the whole earth until the end of the world.

We learned that the church age ended on May 21, 1988, and that the great tribulation will end on May 21, 2011. Although we worked with days when we considered the two parts and the entirety (8400 days) of this great tribulation, notice its total length in years:

23 years (from May 21, 1988 to May 21, 2011)

It is exactly 23 calendar years long, thereby giving us another example of God's use of the number 23 to indicate His judgment and to provide confirmation of these dates.

Following the rapture - which will occur on May 21, 2011 - there will be a period of five months until the end of the world on October 21, 2011. If you count up the number of these days, you will find that it comes to exactly 153. This number may be broken down as:

3 x 3 x 17 (in days, from May 21 to October 21 in 2011)

The number 17 is associated with heaven, and the number 3 with God's purpose. This pattern may, therefore, be a statement of God's purpose to bring His elect to heaven before the final five months of the earth's history so that they may be spared the suffering of that time. Also notice that these 153 days constitute exactly five months according to our modern calendar. Five months is the length of time the flood waters were on the earth during Noah's day. We learn this from Genesis 8:3-4:

> **And the waters returned from off the earth continually: and after the end of the hundred and fifty days the waters were abated. And the ark rested in the seventh month, on the seventeenth day of the month upon the mountains of Ararat.**

It was in the second month on the seventeenth day of the month that "were all the fountains of the great deep broken up, and the windows of heaven were opened." Then, five months later, the ark rested upon the mountains of Ararat in a world greatly transformed by the flood. In these scriptures, God has given us additional proof that our time line is correct, and has shown us that the elect will "rest upon the mountain" in the new earth after judgment day is over. During 5 months of the worlds Judgement

It is interesting to note that the number 153 also appears in the Bible in connection with a catch of fish. In John 21:11, we read:

> **Simon Peter went up, and drew the net to land full of great fishes, an hundred and fifty and three: and for all there were so many, yet was not the net broken.**

The fish in this net all made it to land: the net was not broken as had another net on a different occasion recounted in the Gospels. If your net breaks, you will lose much of your catch - and many in "the catch" during the church age were not true believers. Neither were the 153 fish brought to land in a ship, which in the Bible represents the local congregations. This catch was dragged to land by Simon Peter. These fish appear to represent the true believers who will be saved outside of the local congregations during the period following the end of the church age and continuing until the rapture. These true believers will be spared the 153 days of suffering that will follow May 21, 2011 and continue until October 21 - the very end.

We learned that the second part of the current great tribulation is 6,100 days. This second part began on September 7 1994, and will continue until May 21, 2011. Then there will be a period of 153 days of intense suffering, until the end comes on October 21, 2011. If we consider the total duration of these two time periods together, we get 6,253 days. This number breaks down as:

13 x 13 x 37 (time interval in days from September 7,1994 until October 21, 2011)

This pattern features the number 13 (which we have learned is associated with the end of the world) and the number 37 (associated with God's judgment). This pattern seems to be telling us that God will bring His judgment on the unsaved at the end of the world. Notice that the time interval associated with this pattern uses September 7, 1994 as its starting point. Since we have learned that this date is associated with judgment against the local congregations, this numerical pattern may specifically refer to God's judgment against the local congregations.

There is something else about the October 21 date. We learned that it is linked to the last day of the Feast of Tabernacles and so for that reason alone it is greatly significant. There is, however, something significant about the numbers associated with that date when it is stated in the Jewish ceremonial calendar (also known as the Hebrew calendar): it is the twenty third day of the seventh month. Earlier, we saw that the number 23 is associated with God's wrath or judgment, and that the number 7 is associated with the fulfillment of God's purpose. Based on these numbers, God seems to be indicating that it is His purpose to bring His wrath on the world on October 21 in the year 2011.

For each numerical pattern presented, an attempt has been made to state what God may be teaching us by it. Each statement is an interpretation which may or may not represent God's intention. However, regardless of this, the numbers - as they say - speak for themselves. When we see pattern after pattern, we must know that there couldn't possibly be so many coincidences. We should acknowledge that this is God's work, as Pharaoh's magicians did when Egypt was being subjected to the plagues that forced Israel's release. In Exodus 8:19, we read:

> **Then the magicians said unto Pharaoh, This *is* the finger of God: and Pharaoh's heart was hardened, and he hearkened not unto them; as the LORD had said.**

Despite clear evidence that he was going up against God Himself, and against the advice of those in his court, Pharaoh refused to let the children of Israel go free. Eventually, as a result, Egypt suffered devastation and Pharaoh lost his own life. May we not be like this Pharaoh who chose to ignore God's clear warnings.

Further Confirmation of the Year 2011 as the Final Year

To further confirm that the year 2011 will be the final year of the earth's existence, the Bible provides another proof of a most amazing type. This proof is based on two scriptures - one from the Old Testament and one from the New Testament - and our discovery that 4990 B.C. was the year of Noah's flood.

In the Genesis account of the flood, we read that God gave Noah seven days notice that He was about to bring the rain to destroy all life. Genesis 7:4 states:

> **For yet seven days, and I will cause it to rain upon the earth forty days and forty nights; and every living substance that I have made will I destroy from off the face of the earth.**

There is much more to this scripture than there first appears to be. Of course, there is the obvious statement that God is making to Noah, telling him about the rains that will begin after seven days. As we continue reading this chapter of Genesis, we learn that God did in fact cause the devastating rain to fall just as He had stated;

but there is more to Genesis 7:4 than that.

In order to see what else this scripture may be teaching us, we should look at a very mysterious scripture found in the Book of 2 Peter. In the third chapter of that book, God is telling us about the flood and also about the time of the end when He will destroy the earth by fire. Then, in 2 Peter 3:8, we read something very strange:

> **But, beloved, be not ignorant of this one thing, that one day *is* with the Lord as a thousand years, and a thousand years as one day.**

What could this possibly mean? It is something intended for the elect, not the world at large: the use of the word "beloved" indicates that. It is something important - very important - because it is "one thing" that God wants His elect to know. The repetition also serves to emphasize that this is something very important. The elect are not to be ignorant of this thing, whatever it is. And what is this thing? It is that, with the Lord, a day is as a thousand years and a thousand years as one day. What are we to make of that?

If we compare this scripture with the one we read earlier - the one about the seven days warning that the flood was about to come - we may be very surprised at the hidden meaning we discover. It is this: seven days (that is, seven *long* days of a thousand years each) after the day the flood begins, the earth will be destroyed by fire.

How can we know that this is really what God is telling us? We know because we now understand the Bible's calendar. By working through the calendar, we learned that 4990 B.C. is the date of the flood and that 2011 will be the last year of the earth's existence. Notice that if we begin in the year of the flood and advance the calendar by 7,000 years, we arrive at the year 2011. Here is the calculation (we add 1 because there is no year "0"):

$$-4990 + 7,000 + 1 = 2011.$$

As we saw earlier, the 7,000 year time interval between these two dates also qualifies as a numerical pattern and constitutes a proof that the dates we have discovered are correct.

There is something else amazing to be seen in the flood account when we compare it with the end-time dates we have discovered. In Genesis 7:10-11, we read:

And it came to pass after seven days, that the waters of the flood were upon the earth. In the six hundredth year of Noah's life, in the second month, the seventeenth day of the month, the same day were all the fountains of the great deep broken up, and the windows of heaven were opened.

According to the calendar used in Noah's day, it was on the seventeenth day of the second month that the flood began. We have learned that God is saving a great multitude outside the local congregations, and that this will continue until May 21, 2011. We know that May 21 of 2011 will not be one of the annual feast days, but there is something else noteworthy about that day: according to the Jewish ceremonial calendar, *it is the seventeenth day of the second month.*

These confirmations from Genesis should remove any doubts we may have about the validity of the end-time dates we discovered.

The Final Numerical Pattern: The Doomsday Code

There is one more numerical pattern for us to examine. This may be the most astonishing one of all, because it shows us God's precision in the timing of events in His plan. Before we look at this pattern, however, we should consider a principle God established early in the Bible.

In chapter 41 of Genesis, we read that Pharaoh had two dreams that disturbed him greatly. He dreamt of seven plump cows and seven lean cows, and the lean cows devoured the plump ones. Then he had a similar dream dealing with crops. He dreamt of seven plump ears and seven withered ones, and the withered ones devoured the good ones.

Pharaoh called for Joseph to interpret the dream for him. Joseph told Pharaoh that God had shown Pharaoh what He was about to do. The two dreams were actually one: there would be seven years of plenty, followed by seven years of famine. In Genesis 41:32, we read why the doubling in the dream was important:

And for that the dream was doubled unto Pharaoh twice; *it is* because the thing *is* established by God, and God will shortly bring it to pass.

Let us now examine our last numerical pattern. From our work in the Biblical calendar, we know that the Lord was crucified on April 1 in 33 A.D., on Passover. Then, as one of our four end-time dates, we learned that the rapture will occur on May 21, 2011. Let us calculate the number of days (inclusive) that will have passed since April 1, 33 by the time May 21, 2011 arrives. If we check an encyclopedia, we find that there are approximately 365 days, 5 hours, 48 minutes and 45 seconds in one year. First, let us count the years from April 1, 33 to April 1, 2011. There are 2,011 - 33 years, or 1,978 years. To those years, we will add 30 days of April (because we are counting inclusively) and 21 days in May to get us to May 21, 2011. Converting the 1,978 years into days and rounding the calculations for hours, minutes and seconds to the nearest hundredth of a day gives us:

1,978 x 365 days =   721,970 days
1,978 x 5 hours = 9,890 hours x 1 day/24 hours =   412.08 days
1,978 x 48 minutes = 94,944 minutes
                                 x 1 hour/60 minutes
                                 x 1 day/24 hours =   65.93 days
1,978 x 45 seconds = 89,010 seconds
                                 x 1 minute/60 seconds
                                 x 1 hour/60 minutes
                                 x 1 day/24 hours =   1.03 days

Subtotal   722, 449.04 days

     Dropping the fraction of a day gives us 722,449 days

exactly. Next, we add to the subtotal the 51 days of April and May combined to give us the total:

Subtotal	722,449 days
Days in April and May up to May 21	51 days
Total	722,500 days

What is significant about this number of days? Its significance is in its key factors:

5 x 10 x 17 x 5 x 10 x 17 (inclusive days - time interval from the Crucifixion to May 21, 2011)

What can we say about this numerical pattern? We have learned that the number 5 is associated with the atonement (emphasizing both judgment and salvation), the number 10 with the completeness of whatever is in view, and the number 17 with heaven. When we consider this pattern in view of the time interval from which it came, we may say that the meaning of these numbers is as follows:

It is God's purpose to apply the atoning sacrifice of the Lord Jesus to bring His elect to heaven on May 21, 2011. This will complete God's program of salvation, begun since the foundation of the world. The doubling up (remember Pharaoh's dream about the coming famine) of the numbers indicates that this will definitely happen, and that it will happen shortly.

Take another look at those numbers. This is God's doomsday code. It is pinpointing - to the very day - the beginning of the time of God's great wrath against mankind. The elect won't be here to undergo this terrible period. They will have been raptured on May 21 of 2011. For everyone else, this will be a time of great suffering and hopelessness - a time of death.

Can We Know the Exact Hour Too?

We may be absolutely certain that we have discovered the exact date of the rapture by virtue of the proofs we have seen. Can we know the exact hour too? As we consider this question, let' read what the Lord said to His disciples in Mark 13:32:

> **But of that day and *that* hour knoweth no man, no, no the angels which are in heaven, neither the Son, but the Father.**

Notice that God is linking the day with the hour. He has revealed to us the date when that day (judgment day) begins. Isn't it possible that He has also revealed the hour (the *exact* time, to the hour)? I you check a concordance to see how the word "hour" is used in the Bible, you will find many instances where it means a specific 60 minute period in a day (e.g., the sixth hour, the ninth hour, the tenth hour, etc.). Of course, God knew that we would consider this matter of the exact timing. By writing Mark 13:32 just as He has done, He may possibly be indicating that we should search it out.

Before we consider the possibility that we may discover the exact hour of the rapture, we need to consider something else about Mark 13:32. First, go back a bit in the chapter and read several verses preceding verse 32. Notice that verses 28 to 31 (the parable of the fig tree and two scriptures assuring us that events will come to pass to fulfill all God's words) are sort of tangential to Mark 13:27, which states:

> **And then shall he send his angels, and shall gather together his elect from the four winds, from the uttermost part of the earth to the uttermost part of heaven.**

The "hour" of Mark 13:32 is the time when the elect are gathered from the four winds. Many people read Mark 13:32 and understand it to mean that the timing - the day and the hour of the rapture - cannot possibly be known because not even the Son of God knows it; but the scripture can't have that meaning. The Lord Jesus must know the timing. Remember, it was He who wrote the Bible. He is the Word of God and He is God. It was He who told Daniel to "seal the book" to the time of the end; and it was He who unsealed that book (the Bible), as we read in Revelation 5. It was only after the book was unsealed in these latter days that we learned the timing of God's plan. The Lord Jesus, therefore, must have known the timing; but if that is so, we must search to understand the meaning of the scripture telling us the Son does not know of "that day and that hour."

When we see the word "Son" in Mark 13:32, we immediately think of the Lord Jesus. Surely, that is how the Bible translators understood it, and so it appears in our King James Bible with a capital "S." Notice, however, who else is called a son in the Bible. In 2 Thessalonians 2:3, we read:

> **Let no man deceive you by any means: for *that day shall not come*, except there come a falling away first, and that man of sin be revealed, the son of perdition;**

The word "son" in this scripture refers to Satan and is the same Greek word that appears in Mark 13:32 as "Son." Satan is also called a son in Psalm 89:22:

> **The enemy shall not exact upon him; nor the son of wickedness afflict him.**

We see, then, that the word "son" is also used of Satan. If we understand the "son" in Mark 13:32 to refer to Satan, then we have harmony with everything else we have learned. Satan and his demons may now know the timing of God's salvation plan, but they didn't at the time the Lord Jesus spoke those words recorded in Mark 13:32.

As we consider this matter of the exact timing of the rapture, we find that three of the Gospels indicate that the Lord died at or very near the ninth hour: three o'clock in the afternoon. In Luke 23:44-46, we read:

> **And it was about the sixth hour, and there was a darkness over all the earth until the ninth hour. And the sun was darkened, and the veil of the temple was rent in the midst. And when Jesus had cried with a loud voice, he said, Father, into thy hands I commend my spirit:**

and having said thus, he gave up the ghost.

We know that the Lord is a God of perfect accuracy and precision. If He has decided that there shall be exactly 722,500 days - to the very minute - from the time of His death until the time He takes the elect, then the rapture will occur at exactly the same time of day that the Bible tells us the Lord Jesus "gave up the ghost." The Bible indicates this time to be 3:00 PM Jerusalem time. However, it is not always God's intention that a stated number of days should be *exactly* that number of full 24 hour periods, completed to the very minute. For example, read what the Lord Jesus said to the Pharisees in Matthew 12:39-40:

> **But he answered and said unto them, An evil and adulterous generation seeketh after a sign; and there shall no sign be given to it, but the sign of the prophet Jonas: For as Jonas was three days and three nights in the whale's belly; so shall the Son of man be three days and three nights in the heart of the earth.**

These verses are a prophecy dealing with the duration of the Lord's suffering, beginning before the time He was arrested until the time He arose. We know that He was in the Garden of Gethsemane on a Thursday evening after He had eaten the Passover meal with His disciples. (Remember that a new day begins at sundown, so that the Passover day - which was a Friday in 33 A.D. - began that Thursday evening.) It was in the garden that God began to pour out His wrath on the Lord Jesus, as we read in Luke 22:44:

> **And being in an agony he prayed more earnestly: and his sweat was as it were great drops of blood falling down to the ground.**

From that time until the Lord arose early Sunday morning, we count three nights (Thursday night, Friday night, and Saturday night) and three days or periods of daylight (Friday, Saturday, and part of Sunday).

According to the prophecy, the duration was to be three days and three nights in the heart of the earth. It would seem that this prophecy is telling us that the Lord would be buried for three

Chapter 8 God's Calendar For Our Day

days and three nights. This, however, is not what happened. The Lord was buried late on Friday, and He rose early on Sunday morning. No matter how you count, you can't get three days and three nights from that period of time; but when we realize that being in the "heart of the earth" consisted of two parts - the period of suffering before the Lord's burial and then His time in the tomb - this allows us to understand how the timing of the prophecy applies. The period began on Thursday night in the garden and continued until Sunday morning, and that is how the prophecy of "three days and three nights" was fulfilled.

We have seen that the sign of the prophet Jonah as used by the Lord Jesus did not consist of three complete 24 hour days. Rather, it was a period of time extending from late Thursday to early Sunday. Therefore, we can't say with certainty that the 722,500 inclusive days, counting from the day of the Crucifixion to the day of the rapture, will be completed to the exact minute or hour; and it follows that we cannot declare that judgment day will begin at 3:00 PM.

There is another hour mentioned many times in the Bible. As we read some of the verses where it is used, we find events occurring at this hour - events that seem to be giving us a picture of judgment day. One such example is Acts 16:25-26:

> **And at midnight Paul and Silas prayed, and sang praises unto God: and the prisoners heard them. And suddenly there was a great earthquake, so that the foundations of the prison were shaken: and immediately all the doors were opened, and every one's bands were loosed.**

Based on this and other scriptures where it is used, midnight indeed seems to be a likely possibility for the beginning of judgment day. Notice also that if we consider 3:00 PM to be our benchmark time, midnight occurs 15 hours earlier. Notice how this number may be broken down into two other numbers we have considered:

3 x 5 (time interval in hours from midnight to 3:00 PM)

We have seen that the number 3 is associated with God's purpose, and the number 5 with the atonement. These numbers, therefore, give further support to the possibility that judgment day will begin at midnight (Jerusalem time) on May 21. However, whether judgment day begins immediately after the sun sets over Jerusalem on May 20, at midnight or later that day in the afternoon, it's not so important as this: it will be the day the Lord strikes the earth with a blow that is the beginning of the end of our world.

Chapter 9

God's Wrath

What will happen on May 21, 2011? We have already seen that this is the day God's elect will be taken up to heaven to be with Him: it is the day of the rapture. The Bible, however, tells us more than just that; and what it tells us should strike fear into the heart of anyone who does not believe he is one of God's elect.

The Earthquake

The Bible tells us that there will be an earthquake like no other on the day of the rapture. In Hebrews 12:26-27, we read:

> **Whose voice then shook the earth: but now he hath promised, saying, Yet once more I shake not the earth only, but also heaven. And this** *word***, Yet once more, signifieth the removing of those things that are shaken, as of things that are made, that those things which cannot be shaken may remain.**

This earthquake is also mentioned in the Book of Revelation. Although much of the Book of Revelation must be understood in the spiritual sense, the earthquake it speaks of is mentioned in other books of the Bible in such language that we may be assured that there really will be a terrible earthquake. In Revelation 16:18, we read:

> **And there were voices, and thunders, and lightnings; and there was a great earthquake, such as was not since**

men were upon the earth, so mighty an earthquake, and so great.

The earthquake will cause suffering and death on an unparalleled scale. It will be the most powerful and destructive earthquake of all time. We know the kinds of things that can happen in an earthquake. Therefore, we may expect the following situations to develop when the earthquake strikes and in the aftermath.

Much of the world's population, which is approaching seven billion people at the time of this writing, lives in cities. Many city dwellers live in the great cities of the world characterized by a high population density and very tall buildings. Even the most earthquake-resistant of these tall buildings may not be able to withstand the great earthquake of May 21, 2011. Besides, considering the ages of some cities and the absence of any known earthquake threat reflected in their building codes, it is most likely that relatively few structures have been designed to withstand a great earthquake. These buildings will sustain major damage when the earthquake hits; many will completely collapse. In a single skyscraper, there may be several thousand people; and in a great city, there may be hundreds of skyscrapers.

Of course, there are many areas of the world where there are no skyscrapers. The typical structure in villages of some poorer nations may, for example, be built of mud bricks. Whether built of mud bricks or something else, the less densely populated areas of the world will also suffer great loss of life when their structures come crashing down. We may reasonably assume that very many people, probably numbering in the hundreds of millions, will be killed in the earthquake as a result of collapsing structures. Since May 21 of 2011 will be a Saturday, perhaps a somewhat smaller number will die than on a weekday. (If the earthquake strikes at or soon after the sun sets over Jerusalem on May 20 - for that is when the seventeenth day of the second month begins - it will still be about mid-day on Friday in the United States). In any event, the Bible paints a picture of a world of death. It will be a time of horror.

The air will be badly contaminated in any area where there

is an industrial plant. Many chemicals, some highly corrosive, are stored in great tanks in thousands of locations around the world. These tanks will be badly damaged in the earthquake and spill out whatever is stored in them. When these toxic fluids and gases are exposed to the air, poisonous clouds will result. The severe pollution that results will sicken many persons, some severely enough to kill them.

Many dams, perhaps all of them, will collapse. All around the world, there are some dams of immense size. The catastrophic failure of the great dams will cause very many deaths.

Perhaps tsunamis will kill even more people than collapsing structures. The earthquake may create waves of such size and force that they travel several miles inland - further than ever before in any area which has historically suffered these disasters. Many areas which have never experienced a tsunami in all their recorded history will also be destroyed by a towering wall of water. Additionally, many volcanoes may erupt immediately after the earthquake. Fires, resulting from hot lava flows, are also to be expected. Besides the immediate destruction and deaths caused by these eruptions, there may be also be a global impact on air quality as enormous amounts of volcanic ash are carried high into the atmosphere and fires rage in areas around the volcanoes.

At any moment on any day, there must be millions of people all over the world traveling by car and in other types of motor vehicles. Some of these drivers may be able to pull over safely and stop when the earthquake begins; but for most of them, there won't be enough time. There will be a tremendous number of accidents and fatalities as vehicles crash into each other and into slabs of broken concrete on highways and streets. Trees, light poles, and pieces of various structures will fall and kill many vehicle occupants. Bridges will collapse, dropping many vehicles to the earth and waters below. Underwater and underground tunnels will fail. Occupants of vehicles in these tunnels will drown or be crushed.

Although motor vehicle injuries and deaths should far outnumber those suffered by persons traveling by other methods, it is reasonable to expect that every method of transportation will be devastated. Trains will derail; those traveling underground may very well be crushed as tunnels collapse onto them. Many trains travel along elevated sections of railway - those structures will sustain major damage and many of them are likely to collapse. Ships, especially those near land, will be capsized by tsunamis. Many smaller craft will be tossed around and smashed to bits.

At first thought, it would seem that those traveling by air might be spared the immediate effects of the earthquake; but there may be numerous airplane crashes as well. Such an enormous earthquake may be accompanied by severe electromagnetic disturbances, affecting a plane's instrumentation. There may not be any guidance from the ground either. Air traffic controllers who survive the earthquake will most likely not be able to function. As a result, there will be a high probability of midair collisions. Aircraft flying in bad weather or in mountainous areas of the world will have the added risk of those hazards. Finally, there may be severe down-drafts reaching high into the atmosphere and powerful enough to bring down a plane when the earthquake strikes.

Outside the cities, the natural terrain will undergo severe changes. Some rivers and streams will change course. Many lakes and ponds will be drained, and new bodies of water will form. There will be numerous rock slides. Some of the taller mountain peaks may topple. There will be avalanches and other hazards that crush and drown the people living nearby.

The earthquake will cause an enormous number of injuries as a result of all of these conditions. Many of those injured would most likely be able to survive if they received help. There will be no help in most cases. Many highways and roads will be impassable. Bridges will have collapsed. There will be no electrical power in most locations. Airport landing strips will be damaged. In some areas, helicopters which have not been damaged by collapsing debris may be the only means of reaching those who have been injured or are trapped beneath the rubble; but since fuel supplies will be disrupted, their usefulness will be limited to the

Chapter 9 God's Wrath

fuel they have aboard when the earthquake strikes. Survivors and first responders may also have to cope with aftershocks. Although the Bible appears to tell us only about the one earthquake, we know that there are often a series of smaller earthquakes whenever a great earthquake strikes.

Big cities depend on the continuing arrival of trucks for their food supplies. The trucks will stop coming. Before long, people will begin to starve. Even worse, the vast network of pipes supplying water to the public will be badly damaged. The day of the earthquake may be the last day many people have running water in their homes.

Many fuel pipelines will be broken. Fuel will spill out of these damaged pipelines, contaminating the environment. It may be reasonably expected that in some locations escaped fuel will ignite, resulting in terrible fires.

At any time, there are usually a few wars going on around the planet. Perhaps whatever wars are being fought on May 21, 2011 will simply stop when the earthquake strikes. The nations of the world will most likely be so traumatized by the earthquake that they will simply not be able to wage war. This does not mean, however, that the survivors will be safe from attack by their fellow man. We are not surprised to see images of looting on our television screens these days whenever there is a report of a natural disaster. After the earthquake, this problem will be worse than ever. In addition to looting, there will most likely be a surge in every sort of vile crime. There may be a complete breakdown in law and order.

Some nations of the world rely very heavily on nuclear energy to generate electricity. Radioactive material within reactors and at storage locations elsewhere - often in liquid in vast numbers of drums - will be released when the reactors and the waste storage facilities are damaged. Many thousands may die as a result of radioactive sickness.

There will be other types of plagues and sicknesses as well, resulting from severe pollution and damage to the water and sewage systems. There will be so many dead that it will not be possible for the survivors to bury them all. In Isaiah 34:2-3, we read about that time:

> **For the indignation of the LORD *is* upon all nations, and *his* fury upon all their armies: he hath utterly destroyed them, he hath delivered them to the slaughter. Their slain also shall he cast out, and their stink shall come up out of their carcases, and the mountains shall be melted with their blood.**

Corpses may be above ground for the entire duration of the day of judgment, lasting 153 days, until the earth is burned up later in the year.

There is another scripture that may be telling us something about conditions during the great earthquake. In Isaiah 13:13, we read:

> **Therefore I will shake the heavens, and the earth shall remove out of her place, in the wrath of the LORD of hosts, and in the day of his fierce anger.**

Is God telling us that the earth's orbit will be changed, at least temporarily, when He causes the earthquake? That is a possibility. Once the earthquake has passed, however, normal planetary motion should resume. This is certainly indicated by what we read in Genesis 8:22, which is part of the promise God made to Noah after the flood:

> **While the earth remaineth, seedtime and harvest, and cold and heat, and summer and winter, and day and night shall not cease.**

Off the earth, things will apparently continue as before once the earthquake has passed. Conditions on the earth, however, will be permanently changed.

Dead Men's Bones

We know that the bodies of God's elect will be resurrected

Chapter 9 God's Wrath

and changed into immortal bodies on the day of the rapture. The events following the death of the Lord Jesus give us an indication of this as a picture of the rapture. In Matthew 27:50-53, we read:

> **Jesus, when he had cried again with a loud voice, yielded up the ghost. And, behold, the veil of the temple was rent in twain from the top to the bottom; and the earth did quake, and the rocks rent; And the graves were opened; and many bodies of the saints which slept arose, And came out of the graves after his resurrection, and went into the holy city, and appeared unto many.**

For the non-elect among the dead, it will be an entirely different situation. The Bible paints a very gruesome picture about their remains. In Jeremiah 7:33, we read:

> **And the carcases of this people shall be meat for the fowls of the heaven, and for the beasts of the earth; and none shall fray *them* away.**

There's more about this in the Book of Jeremiah. In Jeremiah 8:1-2, we read the following:

> **At that time, saith the LORD, they shall bring out the bones of the kings of Judah, and the bones of his princes, and the bones of the priests, and the bones of the prophets, and the bones of the inhabitants of Jerusalem, out of their graves: And they shall spread them before the sun, and the moon, and all the host of heaven, whom they have loved, and whom they have served, and after whom they have walked, and whom they have sought, and whom they have worshipped: they shall not be gathered, nor be buried; they shall be for dung upon the face of the earth.**

As we read these verses, we need to remember that Father, Son and Holy Ghost all are God; so when we read "they shall bring out the bones," we should realize that it is God who is doing these things. The Bible indicates that, when the earthquake occurs, graves will be thrown open all around the earth and the remains of millions of dead will be exposed.

Depending on the conditions at death and afterwards, in some cases we would not expect to find even a trace of remains of an individual who has died. It would not be possible, for example to find remains of anyone who was cremated and whose ashes were scattered long ago. In fact, we would ordinarily not expect to find much left of anyone who died thousands of years ago. God however, will be at work to create conditions that are anything but ordinary.

It is not only the elect who will be resurrected. Only the elect will be resurrected to life, but the unsaved will be resurrected too. We know this from the words of the Lord Jesus, as we read in John 5:24-29:

> **Verily, verily, I say unto you, He that heareth my word and believeth on him that sent me, hath everlasting life and shall not come into condemnation; but is passed from death unto life. Verily, verily, I say unto you, The hour is coming, and now is, when the dead shall hear the voice of the Son of God: and they that hear shall live. For as the Father hath life in himself; so hath he given to the Son to have life in himself; And hath given him authority to execute judgment also, because he is the Son of man. Marvel not at this: for the hour is coming, in the which all that are in the graves shall hear his voice, And shall come forth; they that have done good, unto the resurrection of life; and they that have done evil, unto the resurrection of damnation.**

Many scriptures throughout the Bible tell us that death is the fate of the unsaved. We learn that man is like the animals in that way and we learn that in the grave there is no more thought. This is the end of all who have not been saved. Such scriptures help us to understand what is meant by the Lord's words, "resurrection of damnation." Only the elect are resurrected to life. The others will be resurrected as lifeless bodies. They will be corpses, possibly appearing as they would if they had been resurrected to life in their earthly bodies; but there will be no life in them. God may or may not cause them to be animated, to move at least temporarily

Whether or not this happens, the earth will be a place of horror and agony.

What About God's Mercy?

It may be difficult for many people to accept that a merciful God would really allow anyone to suffer so terribly, as many will after the earthquake. After all, wasn't the Lord Jesus - who is God Himself - gentle and kind? Didn't He suffer and die to pay for our sins? Didn't He love children? Only when we read the entire Bible can we begin to understand that judgment day will be horrible indeed. God certainly is merciful, but His mercy will not extend beyond May 21, 2011. This claim may seem to contradict some of the scriptures we read about God's mercy. For instance, in Psalm 100:5, we read:

For the LORD *is* good; his mercy *is* everlasting; and his truth *endureth* to all generations.

How can we reconcile what we know about the hopelessness of anyone who will face judgment day - beginning on May 21, 2011 - with a scripture that tells us that God's mercy is everlasting? In this way: God's mercy is everlasting to His children because He has granted them everlasting life; He will never count their sins against them. In a sense, God's mercy is everlasting even to the non-elect who will have lived and died during all of earth's history before judgment day begins. These unsaved persons will not be punished any more. The death that each of these unsaved persons experiences is the last thing that he or she will ever suffer. God will not require any of them to spend an eternity of suffering in a place called hell, as so many of today's church members believe. This thought should certainly console those of us who have lost loved ones who never showed any evidence of having been saved. On the other hand, God's people must be ever aware that the May 21, 2011 deadline is looming ahead like an enormous iceberg in the path of a ship that will inevitably strike it. That ship is our planet earth.

When we read in the Bible that this is the day of salvation, we need to sense the urgency of seeking God - because the day of salvation is drawing to a close. A statement made by the Lord Jesus to His disciples indicates that the day of salvation would eventually end, as we read in John 9:4:

> **I must work the works of him that sent me, while it is day: the night cometh, when no man can work.**

Until that time, referred to as "night" by the Lord, God will be saving people; true believers must work to get the Gospel out to the world. When that day arrives, there will be no more salvation. The true believers will be gone to be with the Lord, and judgment day will have begun.

When we read and believe the Bible's accounts of the times God executed His judgment in the past, how can we not be terrified at the thought of God's wrath coming again to the earth? If, for example, you have ever read the account of famine in Samaria recorded in 2 Kings, you should not be able to forget it - especially because of the incident recorded in 2 Kings 6:26-31:

> **And as the king of Israel was passing by upon the wall, there cried a woman unto him, saying, Help, my lord, O king. And he said, If the LORD do not help thee, whence shall I help thee? out of the barnfloor, or out of the winepress? And the king said unto her, What aileth thee? And she answered, This woman said unto me, Give thy son, that we may eat him to-day, and we will eat my son to-morrow. So we boiled my son, and did eat him: and I said unto her on the next day, Give thy son, that we may eat him: and she hath hid her son. And it came to pass, when the king heard the words of the woman, that he rent his clothes; and he passed by upon the wall, and the people looked, and, behold, *he had* sackcloth within upon his flesh. Then he said, God do so and more also to me, if the head of Elisha the son of Shaphat shall stand on him this day.**

The king blamed the prophet Elisha; but God caused that famine, and God caused that incident to be recorded. He wanted us to read it and know what happened - to know what He did.

The nation of Israel was God's chosen people, and yet God punished them terribly. Even before they became a nation, they proved to be rebellious; and so God punished them, as we read in Numbers 14:28-30:

> **Say unto them, *As truly as* I live, saith the LORD, as ye have spoken in mine ears, so will I do to you: Your carcases shall fall in this wilderness; and all that were numbered of you, according to your whole number, from twenty years old and upward, which have murmured against me, Doubtless ye shall not come into the land, *concerning* which I sware to make you dwell therein, save Caleb the son of Jephunneh, and Joshua the son of Nun.**

Many times after He had formed the children of Israel into a nation, God sent prophets to warn them of their sins. Incredibly, those warnings are for us as well. God has given us - through the messages recorded by those ancient prophets - warnings of judgment day.

Why Us?

There are many scriptures telling us how terrible God's judgment against the earth will be during the last days. In Isaiah 13:9, for instance, we read:

> **Behold, the day of the LORD cometh, cruel both with wrath and fierce anger, to lay the land desolate: and he shall destroy the sinners thereof out of it.**

The Psalms also give us some scriptures about judgment day. In Psalm 11:5-6, we read:

> **The LORD trieth the righteous: but the wicked and him that loveth violence his soul hateth. Upon the wicked he shall rain snares, fire and brimstone, and an horrible tempest: *this shall be* the portion of their cup.**

Other scriptures may give us a picture of judgment day by telling us about a real historical event. For example, king Jehoshaphat of Judah learned that a great force was coming against his kingdom. In 2 Chronicles 20:3-4, we read how the people reacted to this news:

> **And Jehoshaphat feared, and set himself to seek the LORD, and proclaimed a fast throughout all Judah. And Judah gathered themselves together, to ask *help* of the LORD: even out of all the cities of Judah they came to seek the LORD.**

The king stood in the congregation of Judah, and prayed to God. His beautiful prayer ends in 2 Chronicles 20:12:

> **O our God, wilt thou not judge them? for we have no might against this great company that cometh against us; neither know we what to do: but our eyes *are* upon thee.**

The Lord responded to this prayer through one of His prophets. The people were told that they wouldn't even need to fight, as we read in 2 Chronicles 20:17:

> **Ye shall not *need* to fight in this *battle*: set yourselves, stand ye *still*, and see the salvation of the LORD with you, O Judah and Jerusalem: fear not, nor be dismayed; to-morrow go out against them: for the LORD *will be* with you.**

The next day, Jehoshaphat appointed singers to go out before his army, to sing and praise God. In 2 Chronicles 20:22-24, we read what happened to the invading armies:

> **And when they began to sing and to praise, the LORD set ambushments against the children of Ammon, Moab, and mount Seir, which were come against Judah; and they were smitten. For the children of Ammon and Moab stood up against the inhabitants of mount Seir, utterly to slay and destroy *them*: and when they had made an end of the inhabitants of Seir, every one helped**

to destroy another. **And when Judah came toward the watch tower in the wilderness, they looked unto the multitude, and, behold, they *were* dead bodies fallen to the earth, and none escaped.**

The situation will be similar during judgment day: God's people will be in heaven singing and praising Him while He inflicts His wrath against the people left behind.

We might wonder why God wouldn't just end the world on the day of the rapture. Why would He want to punish the earth between May 21 and October 21 in 2011? Why should He put anyone through 153 days of suffering? Perhaps the answer has to do with man's sinfulness and God's timetable for opening up the Bible to reveal truths hidden in it since it was completed about 1900 years ago.

The account of the great flood is helpful as we struggle to understand judgment day. From the time of Adam until right before the flood, conditions had deteriorated to become what we read about in Genesis 6:12:

> **And God looked upon the earth, and, behold, it was corrupt; for all flesh had corrupted his way upon the earth.**

The Bible tells us that man's sinfulness will once again increase in the last days. This is what we read about in 2 Timothy 3:1-2:

> **This know also, that in the last days perilous times shall come. For men shall be lovers of their own selves, covetous, boasters, proud, blasphemers, disobedient to parents, unthankful, unholy,**

This condition is not only a sign of the last days, it also provides further cause for God to bring judgment on the world.

The evidence of man's sinfulness is all around us, if only our eyes are opened to see it. Sadly, we don't even recognize that so much of what we do is sinful. For instance, if we believe that life on earth began by chance and that mankind evolved from the lower life forms, then we don't believe the word of God. Worse yet, we have made a concerted effort to shut Him out of our daily lives and to eliminate any reference to Him. Yet it is in our day that knowledge of the natural world has increased explosively, and this new knowledge points to God as the Creator. Technological developments, especially over the last several decades, have led to many new findings; and these findings refute the ideas of evolution and an accidental origin of life. For instance, the miniaturization of electronic components has expanded our knowledge of animal behavior. It allows scientists to track animal movements over great distances and to see how they live in their burrows and nests. The theory of evolution just can't explain certain animal characteristics and behaviors (for example, altruism) which are understood today.

Technology has also given scientists new tools and techniques for working at the microscopic level. For example, blood cells with DNA strands were detected in the soft tissue of a Tyrannosaurus Rex fossil. This refutes the belief that the dinosaurs died out millions of years ago. It's a discovery that supports what the Bible indicates: dinosaurs lived much more recently because the earth is only thousands of years old.

Space telescope imagery has given astronomers greater knowledge of the universe. This knowledge includes new information about the expansion of the universe - information that provides reason to doubt scientific claims that the universe is billions of years old, and to support what the Bible tells us about the creation of a very much younger universe. God's word proclaims, in Psalm 19:1:

The heavens declare the glory of God; and the firmament sheweth his handiwork.

Many great scientists of the past acknowledged God's handiwork. Robert Boyle, Isaac Newton, Louis Pasteur, Michael Faraday, Samuel Morse, Charles Babbage and many others were guided by the Biblical truth of God's creation as they pursued their scientific investigations. Today's scientists, on the other hand, seem determined to shut the door on any thinking that allows consideration of the Bible as God's word. No generation has ever been in a better position than ours to acknowledge God's handiwork. God will hold us responsible for failing to do so, as we read in Romans 1:18-21:

> **For the wrath of God is revealed from heaven against all ungodliness and unrighteousness of men, who hold the truth in unrighteousness; Because that which may be known of God is manifest in them; for God hath shewed *it* unto them. For the invisible things of him from the creation of the world are clearly seen, being understood by the things that are made, *even* his eternal power and Godhead; so that they are without excuse: Because that, when they knew God, they glorified *him* not as God, neither were thankful; but became vain in their imaginations, and their foolish heart was darkened.**

Our generation is not only characterized by a refusal to acknowledge God: there are many who actively oppose God's ways and who actually mock God. The Bible tells us that this would happen. In Jude 17-18, we read:

> **But, beloved, remember ye the words which were spoken before of the apostles of our Lord Jesus Christ; How that they told you there should be mockers in the last time, who should walk after their own ungodly lusts.**

If we hold fast to our belief that mankind evolved and life began by chance, then we refuse to acknowledge God; and when we refuse to acknowledge God, we refuse to be guided by His absolute standards of right and wrong. Consequently, we may begin to accept all sorts of sinful behavior in ourselves and in others. An example of this is homosexuality. It is worth noting that of all the types of sinful behavior listed in Romans 1, that chapter lists homosexuality first and expounds on it. Today, many people don't even recognize it as sinful behavior; they just accept it as an alternative lifestyle. Without God to guide us, our idea of moral values becomes warped. In fact, after a time of living without God, a person may not have any moral values at all.

Just as God's word tells us that He has chosen this world's current generation as the one to suffer His end time wrath, we also learn that He has singled out a particular group to be punished even more severely. How do we know this, who are these people, and why has God decided to deal more harshly with them than with others?

On one occasion, the Lord Jesus spoke to His disciples about His return near the end of the world. In Luke 12:40, we read:

Be ye therefore ready also: for the Son of man cometh at an hour when ye think not.

The Lord then proceeded to tell them a parable about servants whose lord was absent. To finish the parable, the Lord Jesus said something very interesting. In Luke 12:46-48, we read:

The lord of that servant will come in a day when he looketh not for *him*, and at an hour when he is not aware, and will cut him in sunder, and will appoint him his portion with the unbelievers. And that servant, which knew his lord's will, and prepared not *himself*, neither did according to his will, shall be beaten with many *stripes*. But he that knew not, and did commit things worthy of stripes, shall be beaten with few *stripes*. For unto whomsoever much is given, of him shall be much required: and to whom men have committed much, of him they will ask the more.

We know that this scripture is concerned with end times because it is telling us something - in the form of a parable - about the return of the Lord Jesus. God is telling us that there will be degrees of punishment: some will be beaten with many stripes, others with few stripes. Of course, we know that God will not punish His elect when judgment day begins; He may have chastised them at some time in their lives, but when judgment day begins they will be taken up to be with the Lord, never to suffer again. We also know that everyone else who is alive when judgment day begins will enter into a period that will last for 153 days. Many will die on the first day, but some will survive until the very end - only to be incinerated on that last day. Some of these people, those who are not saved and are still alive when judgment day begins, will suffer more than others. The Bible tells us who these people are. In Matthew 7:22-23, we read:

> **Many will say to me in that day, Lord, Lord, have we not prophesied in thy name? and in thy name have cast out devils? and in thy name done many wonderful works? And then will I profess unto them, I never knew you: depart from me, ye that work iniquity.**

The persons with whom this parable is concerned believe they have done many wonderful works in the name of the Lord Jesus: they are known to the world as Christians.

The almanac tells us that there are currently about two billion people in the world who consider themselves to be Christians. This, of course, is a very great number. In fact, it is roughly one third of the entire population of the earth. We know that a large number of those who were counted among the two billion belong to local congregations; and we have seen that a works-based gospel, what might be called a do-it-yourself salvation plan, is prevalent in the local congregations; and we have learned that God has commanded His elect to leave the earthly church because Satan is ruling there.

We have not seen great numbers of people leaving their local congregations. If this were to happen, it would be a very newsworthy event; but it seems to be "business as usual" in the local congregations as far as membership numbers are concerned. It is also "business as usual" as far as most of their beliefs are concerned. The local congregations are holding on to their belief that the Lord Jesus will come "as a thief in the night" when He returns. They are trusting in the traditional beliefs of their church - the creeds and doctrines developed by various theologians throughout the church age - rather than trusting in the Bible. These dear people are not concerned at the thought of the Lord's return because they think they are ready for Him. They aren't listening to what the Bible is warning them. In 1 Thessalonians 5:1-4, we read that warning:

> But of the times and the seasons, brethren, ye have no need that I write unto you. For yourselves know perfectly that the day of the Lord so cometh as a thief in the night. For when they shall say, Peace and safety; then sudden destruction cometh upon them, as travail upon a woman with child; and they shall not escape. But ye, brethren, are not in darkness, that that day should overtake you as a thief.

Throughout the church age, the Lord concealed the time-line of His salvation plan. It was not for His disciples to know the date of His return. He had given them their marching orders: to get the Gospel out to the whole world. That was to be the focus of their attention. They thought that the date of the Lord's return was unknowable. This continues to be their thinking, because they are not "watching" as the Lord has commanded; they are not digging into the scriptures, and so they don't know that God has now opened up the scriptures to reveal the date of His return. They aren't concerned that God has already given us precedents for this knowledge: He sent Jonah to the Ninevites to warn them they were facing destruction in 40 days. In Jonah 3:4, we read:

> And Jonah began to enter into the city a day's journey, and he cried, and said, Yet forty days, and Nineveh shall be overthrown.

God's warning to Noah was also a precedent that should make us receptive to the idea that God would tell us the date of His end-time coming. In Genesis 7:4, we read:

> For yet seven days, and I will cause it to rain upon the earth forty days and forty nights; and every living substance that I have made will I destroy from off the face of the earth.

The Bible not only gives us the accounts of Nineveh and the great flood as examples of how God revealed the timing of a coming judgment. In Ecclesiastes 8:5, God actually tells us that we would know both time and judgment:

Whoso keepeth the commandment shall feel no evil thing: and a wise man's heart discerneth both time and judgment.

The local congregations believe they are secure because they are trusting in things they have done - believing that their own works have secured their salvation, and hence they can say "Peace and safety" to each other. They are not concerned about the hour of the Lord's coming because they believe they are ready for it. Besides, they remember what the Lord said in Acts 1:7:

And he said unto them, It is not for you to know the times or the seasons, which the Father hath put in his own power.

This scripture, however, doesn't mean that no man can *ever* know the time. For example, in Romans 3:10, we read:

As it is written, There is none righteous, no, not one:

We all start out under the wrath of God: nobody is righteous; but this scripture isn't saying that no one will ever be righteous. In fact, we know that by God's grace the elect become righteous. As an example, read Matthew 10:41:

He that receiveth a prophet in the name of a prophet shall receive a prophet's reward; and he that receiveth a righteous man in the name of a righteous man shall receive a righteous man's reward.

Our condition of not knowing - on our own - the time of the Lord's coming can be likened to our condition of never achieving righteousness on our own. The warning of 1 Thessalonians 5 to those who refuse to accept the precious warning that God has given us by revealing the time of His coming is that sudden destruction will come upon them. We can already see how the scripture proclaiming that the day of the Lord will come as a thief in the night will be fulfilled in these people: they will be in shock when it happens. God's elect, however, will have heeded the date and so have had a precious opportunity to cry out to God for His mercy.

God is holding those who are alive today - except for the elect - accountable for their sins and worthy of the wrath of end time destruction; but He is especially holding the local congregations responsible because they have the Bible and are not obeying it. In Isaiah 5:20, we read of God's particular anger against these people:

> **Woe unto them that call evil good, and good evil; that put darkness for light, and light for darkness; that put bitter for sweet, and sweet for bitter!**

The Execution, The Shame and The Lost Inheritance

In times past, there was a belief among some peoples of the Pacific Islands that when one of their own died - of any cause whatsoever - somebody was responsible for that person's death. Even if the person died of an illness, old age, starvation, or an accident, the family of the deceased would consult their witch doctor to learn who was responsible so that they might take vengeance on that person. In a sense, they were partially correct in their belief; someone was responsible for that person's death: it was God.

Whenever someone dies, God is responsible - even if the person dies in an epidemic, or in a war, or at the hand of a murderer. If the person was unsaved, then the death was an execution by God. If we recognize that God is in control of all things, then it follows logically that this is the case at the end of our lives. The Bible gives us some clear examples of God's responsibility in this aspect of our lives. In Genesis 38:7, we read:

> **And Er, Judah's firstborn, was wicked in the sight of the LORD; and the LORD slew him.**

And in Exodus 13:15, we find another example:

> **And it came to pass, when Pharaoh would hardly let us go, that the LORD slew all the firstborn in the land of Egypt, both the firstborn of man, and the firstborn of beast: therefore I sacrifice to the LORD all that openeth**

the matrix, being males; but all the firstborn of my children I redeem.

The Bible tells us of other occasions when God has ended the life of an individual. In some of these instances, the person's death should not be considered an execution. God can use a person's death for His own purpose. In 1 Kings 17, the prophet Elijah cried to the Lord to restore the life of a widow's son. Read about this in 1 Kings 17:20-22:

> **And he cried unto the LORD, and said, O LORD my God, hast thou also brought evil upon the widow with whom I sojourn, by slaying her son? And he stretched himself upon the child three times, and cried unto the LORD, and said, O LORD my God, I pray thee, let this child's soul come into him again. And the LORD heard the voice of Elijah; and the soul of the child came into him again, and he revived.**

Elijah knew the child's death was under God's control, and he knew that the Lord could restore that life. The child's death created the circumstances for God to work a mighty miracle through His servant Elijah, and for God to demonstrate His power over life and death to us.

Although the Bible doesn't tell us how the apostle Peter died, sources outside the Bible indicate that he too was crucified. In John 21:18-19, we read what the Lord Jesus told Peter about this:

> **Verily, verily, I say unto thee, When thou wast young, thou girdedst thyself, and walkedst whither thou wouldest: but when thou shalt be old, thou shalt stretch forth thy hands, and another shall gird thee, and carry *thee* whither thou wouldest not. This spake he, signifying by what death he should glorify God. And when he had spoken this, he saith unto him, Follow me.**

The apostle Peter is one of God's elect. His death, like the death of anyone else who has ever died, was under God's control; but Peter's death can't be placed in the same category as the death of an unbeliever. For Peter, death meant exchanging his earthly existence for a heavenly one. The apostle Paul knew of this truth, for in 2 Corinthians 5:8 we read:

> **We are confident, *I say*, and willing rather to be absent from the body, and to be present with the Lord.**

Before the death of any one of God's elect, He gives that person a new, resurrected soul. This is what guarantees life with the Lord after that individual's death on earth. We read a reference to this spiritual life in 2 Corinthians 5:5:

> **Now he that hath wrought us for the selfsame thing *is* God, who also hath given unto us the earnest of the Spirit.**

The unsaved don't have "the earnest of the Spirit." For them, death means the end of life here on earth without any possibility of living again in any way, shape, form or place - ever.

Some may ask how it can be claimed that God is in control of the timing of each person's death when we know that some people decide to commit suicide. This is a reasonable question. We need to always keep two things in mind. First, God knows what is going to happen; He knows, as the Bible tells us, the end from the beginning. Second, God will do whatever needs to be done so that His plans will be accomplished. If God, knowing that an individual has decided to take his own life, wants to delay that person's death until some time in the future, He will not allow that person to commit suicide. God could, for example, change a person's thinking so that he will not try to take his own life; or He could allow the person to make the attempt, but not allow it to succeed; or He could create a situation to prevent the person from even making the attempt. If a person does take his own life, it is only because God has allowed it.

Death marks the end of any suffering for the unsaved; but the Bible tells us that God will shame them before the end of the world, for in the eyes of God they have lived shameful lives. How will this happen? Most of us will eventually have the somber experience of attending a funeral. It is very obvious on such an occasion - and it has undoubtedly always been this way - that the remains of any one who has died are treated with great respect. God, however, will not treat the remains of the unsaved with respect on judgment day. His word tells us that He will shame them by resurrecting their remains and leaving them out in plain sight.

The shame to be associated with the unsaved, who will never know they have been shamed, stands in stark contrast to the shame the Lord Jesus experienced. The Roman soldiers took His clothes, as we read in John 19:23-24:

> **Then the soldiers, when they had crucified Jesus, took his garments, and made four parts, to every soldier a part; and also *his* coat: now the coat was without seam, woven from the top throughout. They said therefore among themselves, Let us not rend it, but cast lots for it, whose it shall be: that the scripture might be fulfilled, which saith, They parted my raiment among them, and for my vesture they did cast lots. These things therefore the soldiers did.**

The Lord was left hanging there, naked, while the soldiers figured how they should divide His clothing among themselves. Of course the Lord Jesus, being God, always knew that He would be subjected to this shame when the day for Him to be crucified finally came. It was yet another aspect of the agony He endured to demonstrate His love for the elect. We read about this in Hebrews 12:2:

> **Looking unto Jesus the author and finisher of *our* faith; who for the joy that was set before him endured the cross, despising the shame, and is set down at the right hand of the throne of God.**

The Lord Jesus endured shame on behalf of God's elect; they have been clothed with His righteousness - symbolized by the coat for which the soldiers cast lots. This casting of lots for the Lord's coat was prophesied about 1,000 years before the Crucifixion, as we read in Psalm 22:17-18:

I may tell all my bones: they look *and* stare upon me. They part my garments among them, and cast lots upon my vesture.

When we follow the Bible's rule of comparing scripture with scripture, we see that the casting of lots has to do with a decision made by God. The elect are clothed with the righteousness of the Lord Jesus, symbolized by His coat in the earlier scripture we read from John 19. God bestowed this righteousness on those *He* chose: the elect can't do anything on their own to achieve that righteousness. Because God sees His elect as being clothed with the Lord's righteousness, these people will not be shamed on judgment day.

There is yet another aspect to the punishment of the unsaved: they lose their inheritance. God created man in His image and likeness for the purpose of sharing with him a glorious life in a new and perfect world for all eternity. Every unsaved person has lost this inheritance, although he or she will never be aware of it.

God gives us a picture of this loss way back in the Book of Genesis. Jacob and Esau were twin brothers, sons of Isaac. One day, when Jacob had boiled some food, Esau returned from the field and was badly in need of a meal. In Genesis 25:29-34, we read what happened:

> **And Jacob sod pottage: and Esau came from the field, and he *was* faint: And Esau said to Jacob, Feed me, I pray thee, with that same red *pottage*; for I *am* faint: therefore was his name called Edom. And Jacob said, Sell me this day thy birthright. And Esau said, Behold, I *am* at the point to die: and what profit shall this birthright do to me? And Jacob said, Swear to me this day; and he sware unto him: and he sold his birthright unto Jacob. Then Jacob gave Esau bread and pottage of lentiles; and he did eat and drink, and rose up, and went his way: thus Esau despised *his* birthright.**

In the Book of Hebrews, there is a reference to this occurrence. In Hebrews 12:16-17, we read:

Chapter 9 God's Wrath

Lest there *be* any fornicator, or profane person, as Esau, who for one morsel of meat sold his birthright. For ye know how that afterward, when he would have inherited the blessing, he was rejected: for he found no place of repentance, though he sought it carefully with tears.

These scriptures also refer to the day when Isaac bestowed his blessings on his sons before he died. Jacob received the choice blessings of his father, although he did it by deception. Years later, Jacob was himself the victim of deception when he was tricked into marrying the sister of the girl he was supposed to marry. Perhaps Jacob understood that this was a repayment from the Lord. Nevertheless, we know that Jacob is one of God's elect, and Esau was a picture of an unsaved person who lost his inheritance. This is the inheritance that the Lord Jesus promised to His disciples. It is an inheritance in a new world, as we read in Matthew 5:5:

Blessed *are* the meek: for they shall inherit the earth.

We also read about this inheritance in Colossians 3:24:

Knowing that of the Lord ye shall receive the reward of the inheritance: for ye serve the Lord Christ.

It is an inheritance reserved in heaven for God's elect. It is an inheritance that will never fade away.

Some Want Punishment to Continue Forever!

Some people have been the victims of terrible cruelty at the hands of one or more individuals. There are undoubtedly many such people who will bear mental and physical scars until their death. In their suffering, they may have consoled themselves with the thought that their tormentors will eventually face the punishment that the churches have traditionally taught as Bible truth: that the unsaved will be condemned to a place called Hell, and there they will suffer in flames for all eternity.

Other people may be consoling themselves with that idea

simply because they are envious of others. Their thinking may be something like this: "I have lead a good moral life all of my life. I have deprived myself of so many things and pleasures that so-and-so regularly enjoys. At least I know that I will be going to heaven, and so-and-so will be paying for his sins for all eternity." Of course, this kind of thinking may be so deeply buried in a person's mind that the individual may not even be aware of it. Nevertheless, if an individual is holding on to that kind of thinking, he can't see what the Bible teaches about the punishment to be inflicted on the unsaved.

A person's failure to accept Biblical truths about judgment may mean that he or she has failed a test that God is using to separate the wheat from the tares. God tells us that He is a merciful God, and the fact that He will eventually end an unsaved person's suffering testifies to that fact.

A time limit on suffering also answers our sense of justice when we consider the fate of an unsaved child. How could a merciful God condemn a little child to an eternity of suffering? Now we know that He won't. Sadly, however, every unsaved child who is alive on May 21, 2011 will enter into that time of judgment, just as the most notorious of unsaved adults will. Undoubtedly, many children perished in the flood waters of Noah's day. The Bible doesn't say that it will any different at the end of time: children will not be exempt from God's wrath.

About the Book of Revelation

We know that there is a lack of concern among many in the local congregations when they hear that the Bible has given us the date for judgment day. They believe the date is unknowable according to the Bible; and besides, they believe they are already prepared for the Lord's return. For some in the local congregations, however, there is another possible explanation for their failure to search the Bible for specific time information: a misunderstanding of the Book of Revelation. Here is the reasoning.

People in the local congregations know that the Lord Jesus

stated that He would come as a thief in the night; but because of their familiarity with the Book of Revelation, they may believe that this warning does not apply to themselves. They may believe that the Book of Revelation reveals that there will be unmistakable signs - in the heavens and on earth - that we are approaching the end of time. They expect to see these signs fulfilled, and so they will know the end is near. Non-Christians, those who have no acquaintance with the Bible, will not know what to make of these signs or events; but those in the local congregations will know.

The trouble with that kind of thinking is that it's based on a literal reading of the Book of Revelation. Much of the book, however, must be understood symbolically or as the language of parables. For example, when we read in the Gospels what the Lord Jesus said about being the good shepherd and knowing His sheep, we know that He wasn't talking about real sheep - He was talking about people. Similarly, when we read about strange creatures and events in the Book of Revelation, we need to realize that some interpretation is required. There are scriptures in the Book of Revelation that *are* to be understood literally, but it is not always obvious how any particular scripture is to be understood.

The Book of Revelation consists largely of a series of strange and terrifying visions God gave to the apostle John. These visions are not in chronological order, thus increasing our difficulty as we struggle to understand; but now that God has revealed the time-line of His salvation plan, we can understand more of this incredible book. Most of it deals with God's end-time judgment. As we read the book, we see that God considers the great army of His elect as an instrument in bringing judgment on the world. Visions relating to this are pictures of armies clashing in a great battle. This is the book that has given us the well known term "the battle of Armageddon," as we read in Revelation 16:14-16:

> **For they are the spirits of devils, working miracles, *which* go forth unto the kings of the earth and of the whole world, to gather them to the battle of that great day of God Almighty. Behold, I come as a thief. Blessed**

is he that watcheth, and keepeth his garments, lest he walk naked, and they see his shame. And he gathered them together into a place called in the Hebrew tongue Armageddon.

This will be a battle in which there won't be any shots fired. To help us understand these scriptures, Revelation 9:16-18 should also be considered:

> And the number of the army of the horsemen *were* two hundred thousand thousand: and I heard the number of them. And thus I saw the horses in the vision, and them that sat on them, having breastplates of fire, and of jacinth, and brimstone: and the heads of the horses *were* as the heads of lions; and out of their mouths issued fire and smoke and brimstone. By these three was the third part of men killed, by the fire, and by the smoke, and by the brimstone, which issued out of their mouths.

Is this vast army of 200 million under Satan's control, as their horses spew out fire and smoke and brimstone to kill the third part of men? Let's first read Genesis 19:24:

> Then the LORD rained upon Sodom and upon Gomorrah brimstone and fire from the LORD out of heaven;

It turns out that fire and brimstone are used to represent destruction that *God* brings to the wicked. The 200 million horsemen must be the elect of God. We know, however, that God's elect won't even be here after the earthquake. And what about that third part of men who are killed? What is going on here? God seems to be telling us that the rapture of the elect will be a judgment against the unsaved. Even though the elect will no longer be on the earth, it's as if they are condemning those left behind. It's as if God's elect are attacking the unsaved with fire, and smoke and brimstone. As for the third part of men, it may be that God is using that term to refer to the world's approximately two billion Christians. This group of people is about a third of the world's current population, and are found mainly in the local congregations.

Chapter 9 God's Wrath

In the Gospels, there are scriptures to support this understanding of the elect as an army of God. For example, in Matthew 12:41, we read:

> **The men of Nineveh shall rise in judgment with this generation, and shall condemn it: because they repented at the preaching of Jonas; and, behold, a greater than Jonas *is* here.**

The people of Nineveh, whom God saved after they heard Jonah's warning, will be in that great "army" we read about in the Book of Revelation. The local congregations will see them among millions of others ascending in the rapture and know that they themselves have been left behind. This will be the time of "weeping and gnashing of teeth" of which the Lord Jesus warned us. We read about this in Luke 13:25-28:

> **When once the master of the house is risen up, and hath shut to the door, and ye begin to stand without, and to knock at the door, saying, Lord, Lord, open unto us; and he shall answer and say unto you, I know you not whence ye are: Then shall ye begin to say, We have eaten and drunk in thy presence, and thou hast taught in our streets. But he shall say, I tell you, I know you not whence ye are; depart from me, all *ye* workers of iniquity. There shall be weeping and gnashing of teeth, when ye shall see Abraham, and Isaac, and Jacob, and all the prophets, in the kingdom of God, and you *yourselves* thrust out.**

Just as God shut the door to the ark on the seventeenth day of the second month and opened the "windows of heaven" in 4990 B.C., he will shut the door again 7,000 years later - on May 21, 2011, being the seventeenth day of the second month according to the ceremonial calendar - to any possibility for salvation.

There is something else interesting about the scripture we read earlier from the Book of Revelation - the one telling us about the 200 million. Notice that after the number is given as two hundred thousand thousand, we read: "and I heard the number of them." It's almost as if God is giving us the number twice. Why should God be calling our attention to this number? It is very possible that God is revealing to us here the actual total number of His elect.

When we read the Bible, we find that there is little hope that many people were saved throughout history. God dealt with relatively few persons until He established the nation of Israel; and afterwards, throughout Israel's history, we don't find many people being saved. For example, in 1 Kings 19 God told Elijah that He had 7,000 left in Israel. Of course, that was after others had been killed; so we don't know how many other servants of the Lord there had been. Elijah thought he was the only one left when he spoke with God, as we read in 1 Kings 19:14:

> **And he said, I have been very jealous for the LORD God of hosts: because the children of Israel have forsaken thy covenant, thrown down thine altars, and slain thy prophets with the sword; and I, *even* I only, am left; and they seek my life, to take it away.**

God then told Elijah there were 7,000 left. Assuming that Israel and the other nations of that day each had a population of one to two million people, that is a very small percentage. The biggest number we find for people saved at any one time must be the 120,000 people of Nineveh. The only big numbers we find in the church age are the thousands who were saved when it began; and we know that the church age was the "small harvest."

Perhaps God saved several million people throughout all history - we just don't know; but we do know that in our day a great multitude is being saved. Most, but not all, of the world's recognized two billion Christians are in local congregations; and most, but not all, of the possible 200 million elect are alive and being saved today. Interestingly, these two groups relate to each other in approximately a ten-to-one ratio. What we read in Amos 5:2-3 gives further support for us to understand the 200 million as a literal number:

The virgin of Israel is fallen; she shall no more rise: she is forsaken upon her land; *there is* **none to raise her up. For thus saith the Lord GOD; The city that went out** *by* **a thousand shall leave an hundred, and that which went forth** *by* **an hundred shall leave ten, to the house of Israel.**

Although 200 million isn't a very great number compared to the population of recognized Christians in the world today, we at least have hope in knowing that it is in our day that God is saving more people than ever before in history.

Chapter 10

Now What?

This book's warning is surely the last thing the world wants to hear. Even those whose lives are extremely difficult have hope for a better tomorrow, a hope to enjoy some of the best things this world has to offer. After all, this world is exciting, interesting, and beautiful. This is where we have our friends and loved ones, and where sights and sounds bring to mind our treasured memories. Who can bear to hear that the brief time until May 21, 2011 is all we have left for a normal life, and that October 21, 2011 is the day our doomed planet will be destroyed? Yet this is exactly what God's word is telling us today. It is a warning we dare not ignore.

The Bible is God's message to all mankind. Even an unsaved person can read and understand that the Bible tells us God created the heavens and the earth, that the children of Israel were freed from slavery in Egypt, that the Lord Jesus was crucified, and many other things. An unsaved person can't really understand salvation, for example, because only God can open a person's mind to understand the spiritual aspects of the Bible; but the unsaved can understand that God is a merciful God. Read, for example, what God reveals about Himself in 2 Chronicles 30:9:

> **For if ye turn again unto the LORD, your brethren and your children *shall find* compassion before them that lead them captive, so that they shall come again into this land: for the LORD your God *is* gracious and merciful, and will not turn away *his* face from you, if ye return unto him.**

We have learned that we all need that mercy because we can't obey God's commands in a way that can ever earn us salvation.

Why Should We Seek God's Mercy?

God's laws apply to everyone, and God's laws have always benefited mankind. Just think how much better life would have been for hundreds of thousands of unsaved people throughout history if they had only followed the Bible's command against - for example - drunkenness. Now, with the looming danger of these last days hanging over all humanity, it is more important than ever that people listen to the Bible's warnings and turn to God. We all stand guilty before God's law, and that law imposes a severe penalty when we break it. God, in His mercy, has paid the penalty for some. Those for whom He paid the penalty will be called children of God; but what about everyone else? Perhaps God will be merciful even to the unsaved on judgment day by cutting short their suffering - if they cry out to God for mercy before judgment day begins. It would be far better to die quickly than to live out the entire 153 days of hell on earth. God may grant to those non-elect who ask Him for mercy a quick and possibly even painless death. There is no doubt, however, that some of the unsaved will suffer terribly. God is especially angry at the local congregations. Will their members who refuse to heed the call to leave the congregations be targeted for especially painful physical suffering? God's word warns us of plagues, as we read in Revelation 18:4:

> **And I heard another voice from heaven, saying, Come out of her, my people, that ye be not partakers of her sins, and that ye receive not of her plagues.**

Perhaps God's particular wrath against the local congregations will be the extreme mental anguish they will experience when they see God's elect going up to heaven and they themselves left behind. Whatever form of additional punishment God has in store for the local congregations, it will undoubtedly be severe. Unsaved Christian parents of young children who are left behind with their children will certainly dread each new day after May 21, 2011. The Bible doesn't say anything about God having a separate plan for young children. It will be the same for those unsaved children as it was for the children in Noah's day: they suffered and perished with everyone else. If your children haven't been saved by the time judgment day begins, they will enter into that terrible time along with all the unsaved adults of the world. That is why it is so important for parents to leave their local congregations now. It is only outside local congregations that God is saving anyone through His word, and it is only to those outside the local congregations that God may be merciful.

True Believers Have a Grave Responsibility

In these last months of the earth's existence, God has given His elect the task of warning the world that judgment day is almost here. This is the same responsibility God gave to the prophet Ezekiel, as we read in Ezekiel 33:2-7:

> **Son of man, speak to the children of thy people, and say unto them, When I bring the sword upon a land, if the people of the land take a man of their coasts, and set him for their watchman: If when he seeth the sword come upon the land, he blow the trumpet, and warn the people; Then whosoever heareth the sound of the trumpet, and taketh not warning; if the sword come, and take him away, his blood shall be upon his own head. He heard the sound of the trumpet, and took not warning; his blood shall be upon him. But he that taketh warning shall deliver his soul. But if the watchman see the sword come, and blow not the trumpet, and the people be not warned; if the sword come, and take *any* person from among them, he is taken away in his iniquity; but his blood will I require at the watchman's hand. So thou, O son of man, I have set thee a watchman unto the house of Israel; therefore thou shalt hear the word at my mouth, and warn them from me.**

The world will not hear the warning from the source they would expect. Any non-Christian would certainly expect a warning from the God of the Bible to come from a group that publicly acknowledges the Bible to be the word of God and appears to be

His people. No, the warning will not come from the local congregations. Notice that God told Ezekiel that he "shalt hear the word at my mouth, and warn them from me." The local congregations aren't listening to the Bible, and so they can't hear.

The warning will only come from sources outside the local congregations. That's how it was in the days of Micaiah - he was an outsider, and only he warned against war with Syria. In 1 Kings 22:6, we read:

> **Then the king of Israel gathered the prophets together, about four hundred men, and said unto them, Shall I go against Ramoth-gilead to battle, or shall I forbear? And they said, Go up; for the Lord shall deliver *it* into the hand of the king.**

The prophet Micaiah was also summoned to appear before the king and to prophesy. He alone warned against going to battle. In 1 Kings 22:17, we read what Micaiah said in his prophecy:

> **And he said, I saw all Israel scattered upon the hills, as sheep that have not a shepherd: and the LORD said, These have no master: let them return every man to his house in peace.**

Micaiah had yet more to say as he explained what the Lord had revealed to him. In 1 Kings 22:19-23, we read:

> **And he said, Hear thou therefore the word of the LORD: I saw the LORD sitting on his throne, and all the host of heaven standing by him on his right hand and on his left. And the LORD said, Who shall persuade Ahab, that he may go up and fall at Ramoth-gilead? And one said on this manner, and another said on that manner. And there came forth a spirit, and stood before the LORD, and said, I will persuade him. And the LORD said unto him, Wherewith? And he said, I will go forth, and I will be a lying spirit in the mouth of all his prophets. And he said, Thou shalt persuade *him*, and prevail also: go forth, and do so. Now therefore, behold, the LORD hath put a lying spirit in the mouth of all**

these thy prophets, and the LORD hath spoken evil concerning thee.

They should have listened to Micaiah, for he alone spoke the truth. In these last days, we know that God is preparing the world for judgment and that God Himself has deluded the local congregations into erroneous thinking. The warning of 2 Thessalonians 2:10-12 certainly applies to our day:

And with all deceivableness of unrighteousness in them that perish; because they received not the love of the truth, that they might be saved. And for this cause God shall send them strong delusion, that they should believe a lie: That they all might be damned who believed not the truth, but had pleasure in unrighteousness.

Because of this delusion, it may be that those outside the local congregations - even those who have no acquaintance whatsoever with the teachings of the Bible - are more likely to seek God's mercy and to receive it when they finally do hear God's word. These people - the world's Buddhists, Hindus, Muslims and followers of other religions - constitute most of the world's population. They need to recognize that only the Bible is God's word.

The Bible went out to all the world during the church age. Many millions of people have read it. Could any one of them have possibly dreamed that God had hidden within its pages the exact date for the end of the world? Much more than that: He hid the dates for the major turning points in His salvation plan; and, to top it all off, He hid mathematical proof that these dates are correct. God has revealed all of these things to us now, because we are living at the time of the end. This information, and the related warnings, are now going out all over the world by radio and other electronic means.

If you have read this book carefully, you have seen the proof that these things are so. You may have a fatalistic attitude about all of this, the kind of thinking that says you can't do anything to change your situation so you may as well "eat, drink and be merry." On the other hand, you need to recognize that you can't be wiser than God; and God tells us to seek Him and His mercy.

What Should An Unsaved Person Do Now?

The short answer to that question is: pray - especially for God's mercy - and read the Bible. Get yourself a good Bible preferably a King James Version of the Bible. If you must use an NIV version of the Bible, then read that. In the event that circumstances prevent you from reading the Bible regularly, you may be able to find a convenient source to listen to the Bible being read. It's no longer necessary to find someone willing to visit you and read to you, although that's a nice option if you have it. These days radio, television, the Internet, and CD recordings are all possible sources for you to listen to Bible readings. Either by listening or by reading the Bible yourself, you are placing yourself in the environment in which God can save you.

If you have read this book carefully, you have followed the path by which the key dates in God's salvation plan were discovered. It was a long path, with many steps. You have also seen numerous mathematical proofs confirming that those dates are correct. You must realize that those proofs validate each step in the path. Therefore, the proofs establish the following beyond any doubt:

1. *The Bible is the Word of God.* Many different scriptures were used to discover the dates and the proofs. Each of those scriptures is like an essential clue in an incredible mystery. The mystery could only be solved when we had all the clues, and the earliest clues appeared several hundred years before the last ones. Therefore, only God could have been responsible for those dates and the related proofs.

2. *You can only know the truth if you accept the Bible as the Word of God, excluding all else.* There can be no room for you to accept any other writing as being holy. There is no

Chapter 10 Now What?

other source of truth, whether they be writings, visions, customs and traditions, individuals or groups of people - even those among whom genuine miracles are being performed.

3. *You can't read the Bible the same way you read everything else.* God made the Bible extremely difficult to understand. We must compare scripture with scripture. We must let the Bible itself define the meaning of the words as they were used in the original languages; and we must pray for understanding.

4. *The Lord Jesus is God.* He is the Redeemer and the Messiah promised since the time of ancient Israel. If you are rejecting Him, you're in great danger.

5. *The earth is only a little more than 13,000 years old.* Relatively few people accept this "young earth hypothesis," choosing instead to believe that the earth is several billions of years old.

6. *Man did not evolve.* An important premise in the theory of evolution is that there were millions of years during which small and gradual changes transformed earlier living organisms into man. When the premise has been proved false, the entire argument for evolution collapses. The Bible tells us that the earth is very much younger than the evolutionists require it to be as part of their theory, and that man was created as man.

7. *Satan is real and powerful.* Throughout history, he has deceived man. He may have been responsible for the establishment of one or more of the world's major religions. He could have appeared as an angel of God, worked miracles, and dictated one or more books that are today considered holy writings by many people. Today he is even ruling in the Christian churches.

8. *The church age has ended.* If you're in a local congregation get out now. Satan is ruling there.

9. *You cannot save yourself.* There is no formula or ritual; there is nothing you can say or do or think that can save you. It is only by the grace of God that anyone can be saved.

10. *Judgment day is extremely close and it will be horrible, so the matter of seeking God's mercy is urgent.* Judgment day begins on May 21, 2011. From then on, God will no longer be merciful to save anyone.

These items suggest that there is plenty for you to pray about. You can pray for help in understanding God's word, and being obedient to it; you can pray for forgiveness for all the times you have broken God's laws; you can pray that God will deliver you from any sin that has entrapped you; you can pray to Him for relief in a painful situation; you can thank God for your life, and every good gift He has bestowed on you; you can praise Him for the incredible beauty of His creation; you can ask God to save you and your loved ones; and you can thank Him for giving you the opportunity to seek His mercy in these last days of the earth's existence.

The Lord Jesus has given us a wonderful example of prayer in Matthew 6. It begins in Matthew 6:9, where we read:

After this manner therefore pray ye: Our Father which art in heaven, Hallowed be thy name.

Notice that we are to address our prayers to the Father: not to the Virgin Mary, or any other person. We must pray to the God of the Bible *only*, not to Him *and* someone else - "just in case." There are some other things we should try to remember as we pray.

Prayer is half of a conversation with God. He speaks to us through the Bible, and we talk to Him through prayer.

Inevitably, we will be praying for help with our troubles in life. Our prayers may or may not result in a noticeable change in

Chapter 10 Now What?

our circumstances. If it is God's will, He can deliver us out of an extremely difficult, even impossible, situation. The prayer doesn't have any power in and of itself; the power rests in God. Whatsoever you do ask in prayer, ask it in Jesus' name, for this too is according to the Bible. As we ask God for help to relieve us from our troubles and to strengthen us to bear them, we should also remember to thank Him for His blessings and to praise Him for everything. We shouldn't let our petitions become the only reason for prayer.

We need to have the right attitude in our heart when we pray. We are God's creatures, going to the Creator. We are totally dependent on Him. God is so far above man that we can't understand it, and He has no obligation to hear our prayers.

God may use our prayers to change us. If there is a change, however, it might not immediately be to our liking. God might put us through very difficult times in order to change us. If we have prayed for salvation - that is, for eternal life - then we certainly need to be changed.

Our prayer should not be something we have memorized and say over and over again during a single occasion of prayer, or from one day to the next. We should use our own words - and God's words that we have read in the Bible - when we pray. God inspired what we read in Psalm 119:11 for our instruction and as an example for us:

Thy word have I hid in mine heart, that I might not sin against thee.

We too should "hide God's words in our heart," so that when we pray our prayers are shaped by God's own words. Remember, prayer is your conversation with God. What kind of a conversation would it be if you said exactly the same thing in exactly the same words every time you met someone you know?

You can pray anytime and anyplace. Your prayer may be very brief or very long. It is a good idea to try to set aside at least some time each day when you are by yourself so that you can pray for a while without interruption or distraction.

God originated language, and He understands all the world's languages and dialects. No matter which of these you use, you may pray to Him and seek His mercy. May He grant it to you and to us all.

Appendix 1
A Direct Derivation of the Rapture Date

Although the process of establishing the four end time dates is long and complex, there is a relatively direct path by which we may arrive at the date of the rapture. We have seen that we can mark the passage of time and learn the timetable of the earth's existence when we correctly understand the scriptures concerning the life spans of the calendar patriarchs. In this Appendix, we will use scriptures pertaining to the calendar patriarchs who lived from Creation until the time of the flood.

In Genesis 2:2, we read:

And on the seventh day God ended his work which he had made; and he rested on the seventh day from all his work which he had made.

The seventh day of this scripture marked the end of the first week of time. It was the week in which God created the material universe, including man. The first man, Adam, was created on the sixth day of that very first week. We begin our count at year "0" - the Creation.

The Creation: Biblical Year 0

In Genesis 5:3, we read:

And Adam lived an hundred and thirty years, and begat *a son* in his own likeness, after his image; and called his name Seth:

Seth was Adam's son, and so we advance the Biblical calendar by

130 years to reflect Adam's age when Seth was born:

                                Biblical Year   0
Number of Adam's years to Seth's birth   + 130
                                Biblical Year   130

After Seth comes Enos, as we read in Genesis 5:6:

And Seth lived an hundred and five years, and begat Enos:

Again, we have a father-son relationship here. We advance the Biblical calendar by 105 years to reach the year Enos was born:

                                Biblical Year   130
Number of Seth's years to Enos' birth   + 105
                                Biblical Year   235

The next calendar patriarch is Cainan, as we read in Genesis 5:9:

And Enos lived ninety years, and begat Cainan:

This Cainan was not the son of Enos. Rather, he was a distant descendant who was born the year Enos died. Amazingly, we need to go to an entirely different part of the Bible - Luke Chapter 3 - to understand this. (Comparing Luke 3:35-36 with Genesis 11:12, we find an additional name has been inserted between Arphaxad and Salah. In order for both genealogies to be true, it follows that Salah was not Arphaxad's son, but rather a descendant of a later generation.) Until we reach Lamech and Noah, who were father and son, we treat the calendar patriarchs who follow Cainan in the

Appendix 1 A Direct Derivation of the Rapture Date

same manner as we do for Cainan - by adding lifespan to lifespan. Enos' period continued for 905 years, as we read in Genesis 5:11:

> **And all the days of Enos were nine hundred and five years: and he died.**

Advancing the Biblical calender accordingly for Enos, we get:

$$\begin{array}{lr} \text{Biblical Year} & 235 \\ \text{Lifespan of Enos} & +\,905 \\ \hline \text{Biblical Year} & 1140 \end{array}$$

In Genesis 5:12, we read the name of the calendar patriarch who followed Cainan:

> **And Cainan lived seventy years, and begat Mahalaleel:**

We find the lifespan of Cainan in Genesis 5:14:

> **And all the days of Cainan were nine hundred and ten years: and he died.**

And so we advance the Biblical calendar for Cainan by 910 years:

$$\begin{array}{lr} \text{Biblical Year} & 1140 \\ \text{Lifespan of Cainan} & +\,910 \\ \hline \text{Biblical Year} & 2050 \end{array}$$

After Mahalaleel, the next calendar patriarch is Jared, as we read in Genesis 5:15:

And Mahalaleel lived sixty and five years, and begat Jared:

Then, in Genesis 5:17, we read the lifespan of Mahalaleel:

And all the days of Mahalaleel were eight hundred ninety and five years: and he died.

Therefore, we add Mahalaleel's 895 years to our running total:

$$\begin{array}{lr} \text{Biblical Year} & 2050 \\ \text{Lifespan of Mahalaleel} & +\,895 \\ \hline \text{Biblical Year} & 2945 \end{array}$$

In Genesis 5:18, we learn that Enoch follows Jared:

And Jared lived an hundred sixty and two years, and he begat Enoch:

And in Genesis 5:20, we find how long Jared lived:

And all the days of Jared were nine hundred sixty and two years: and he died.

We advance the Biblical calendar by 962 years for Jared:

$$\begin{array}{lr} \text{Biblical Year} & 2945 \\ \text{Lifespan of Jared} & +\,962 \\ \hline \text{Biblical Year} & 3907 \end{array}$$

We learn from Genesis 5:21 that Methuselah - the man who lived longer than anyone else whose lifespan is given in the Bible - comes after Enoch as the next calendar patriarch:

And Enoch lived sixty and five years, and begat Methuselah:

Then, in Genesis 5:23, we learn the lifespan of Enoch:

And all the days of Enoch were three hundred sixty and five years:

Adding Enoch's number to the running total gives us:

$$\begin{array}{ll} \text{Biblical Year} & 3907 \\ \text{Lifespan of Enoch} & +365 \\ \hline \text{Biblical Year} & 4272 \end{array}$$

Genesis 5:25 tells us that Lamech followed Methuselah as the next calendar patriarch:

And Methuselah lived an hundred eighty and seven years, and begat Lamech:

In Genesis 5:27, we learn how long Methuselah lived:

And all the days of Methuselah were nine hundred sixty and nine years: and he died.

And so we advance the Biblical calendar by Methuselah's amazing lifespan:

$$\begin{array}{ll} \text{Biblical Year} & 4272 \\ \text{Lifespan of Methuselah} & +969 \\ \hline \text{Biblical Year} & 5241 \end{array}$$

The next verse, Genesis 5:28, tells us Lamech's age when Noah was born:

And Lamech lived an hundred eighty and two years, and begat a son:

If we read the verse that follows this one, we learn that this son was Noah. Based on the method we have been using, one might expect that we should add Lamech's lifespan to the Biblical calendar as our next step. The Bible gives us the information we need to advance the Biblical calendar, but - as we shall see - the information is presented differently for Lamech and Noah than it is for the earlier calendar patriarchs. First, let's advance the Biblical calendar to reflect the year of Noah's birth:

Biblical Year   5241
Number of Lamech's years to Noah's birth   + 182

Biblical Year   5423

Genesis 7:6 gives us Noah's age at the time of the flood:

And Noah *was* six hundred years old when the flood of waters was upon the earth.

Advancing the Biblical calendar accordingly brings us to the year of the flood:

Biblical Year   5423
Noah's age when the flood began   + 600

Biblical Year   6023

So we see that the flood occurred in the year 6023 according to the

Appendix 1 A Direct Derivation of the Rapture Date

Biblical calendar.

Recall from Chapter 6 that when we synchronized the Biblical calendar with the modern calendar, we learned that the Biblical year 10,416 corresponded to the year 597 B.C. in our modern calendar. Now,

$$10,416 - 6,023 = 4,393$$

So we see that the flood occurred 4,393 years earlier than our benchmark year of 10,416 in the Biblical calendar. We need only to subtract 4,393 years from 597 B.C. to know the date of the flood according to the modern calendar.

$$- 597 - 4,393 = - 4,990$$

And so we know that 4990 B.C. is the date of the flood. Next, we use the clues that God has provided to tell us how long after the flood it would be until the end of time. In Genesis 7:4, we read:

> **For yet seven days, and I will cause it to rain upon the earth forty days and forty nights; and every living substance that I have made will I destroy from off the face of the earth.**

The next rain of global destruction from God will not be water; it will be of the type we read about in Psalm 11:6, and it will mark the end of the world:

> **Upon the wicked he shall rain snares, fire and brimstone, and an horrible tempest: *this shall be* the portion of their cup.**

Next, we go to 2 Peter 3:8:

> **But, beloved, be not ignorant of this one thing, that one day *is* with the Lord as a thousand years, and a**

thousand years as one day.

Here, God is equating a day with a thousand years; but God' warning to Noah in Genesis 7:4 was a warning of destruction i seven days. Advancing the calendar by 7,000 years - a thousan years for each day of God's warning to Noah - from 4990 B.C brings us to:

$$-4{,}990 + 7{,}000 + 1 = 2{,}011$$

which is the year 2011 (note that we add 1 to account for the uni that is used up as we go through 0, because there is no year "0"). I this manner we arrive at 2011 as the year to which God is directing us.

Just as God warned Noah that there were only seven day remaining for anyone who would be saved from the flood to ge into the ark, God has now opened our eyes to understand thi warning about the end of the world:

Seven thousand years after the flood, God's final judgment wil come upon the earth. The Lord Jesus is our "Ark" and our only means of escaping this judgment.

When we examined the various confirmations that the year 2011 will be the final year of the earth's existence, we found that May 21 of that year - according to the calendar used by Jews today - wil be the seventeenth day of the second month. That is also the day indicated by the Bible, as we read in Genesis 7:11-12:

> **In the six hundredth year of Noah's life, in the secon month, the seventeenth day of the month, the same day were all the fountains of the great deep broken up, an the windows of heaven were opened. And the rain wa upon the earth forty days and forty nights.**

It will be just as it was in Noah's day. Noah entered the ark seven days before the flood began, and the Lord shut him in as we read in

Genesis 7:16:

And they that went in, went in male and female of all flesh, as God had commanded him: and the LORD shut him in.

On May 21, 2011, God will shut the door to any possibility of saving anyone else.

Appendix 2
Turning Back Our Calendar To The Crucifixion

You may recall that earlier in this book exact dates, even to the month and day, were stated for several ancient Biblical events. Those events occurred on annual feast days according to the Jewish calendar. It was stated then that the dates for those feast days were determined by calculations, although the calculations were not given. It's important to realize that the mathematical proofs we presented confirm the correctness of those dates. The proofs are based on time intervals between pairs of dates from the overall structure of dates that God has revealed to us. The proofs revealed a date structure so precise that no date can be shifted by even a single day, and for this reason we can have complete confidence that the rapture will occur on May 21, 2011. Nevertheless, it would be interesting to investigate how those dates may be determined by calculations. Let's try to do this for one of those dates: the date of the Lord's death on the Passover in 33 A.D. Our timepieces for such calculations are the sun and the moon, and our rule book is the Bible.

An Anniversary Based on the Tropical Year

To begin our investigation of time, let's conduct a "thought experiment" - as Albert Einstein would have called it. We're not as smart as he was, so our thought experiment will be very simple. Let's suppose that back in the year 2000, on a particular day and at an exact time, a very memorable event occurred in your life. For our purpose, we'll pick March 20 at 7:34 A.M. Let's say that you were so happy about this event in your life that you decided to celebrate it every year from then on; but you don't want to just celebrate it according to the calendar - you want to celebrate near the exact moment as determined by the earth's rotation around the sun. We have seen that the length of a year (note that we're using a tropical year, on which the return of the seasons depends) is 365 days, 5 hours, 48 minutes, and 45 seconds. Let's say that you use 365 days, 5 hours, and 49 minutes in order to determine each anniversary (rounding off the 45 seconds to the next minute). These are the times you would calculate for your anniversary each year:

TEN SOLAR ANNIVERSARIES OF
MARCH 20, 2000 AT 7:34 A.M.

Year	Day	Time
2001	March 20	1:23 P.M.
2002	March 20	7:12 P.M.
2003	March 21	1:01 A.M.
2004	March 20	6:50 A.M.
2005	March 20	12:39 P.M.
2006	March 20	6:28 P.M.
2007	March 21	12:17 A.M.
2008	March 20	6:06 A.M.
2009	March 20	11:55 A.M.
2010	March 20	5:44 P.M.

By adding five hours and 49 minutes to the time each year, the time for the next anniversary is determined. Notice that when you do this a few times, the anniversary date advances to the next day; but soon a leap year comes along and brings the anniversary date back by one day. Now think about the reverse process. If you know the date and time for the anniversary in the current year, you could work backwards to determine the date and time of the original event.

The Consistency of Spring

Although it is very unlikely that anything special occurred in your life at exactly 7:34 A.M. on March 20 in the year 2000, that particular date and time do have significance: that was the exact time (using Greenwich Mean Time or GMT) that spring arrived at the turn of this century. We know that spring arrives pretty much at the same time each year - that is, from one tropical year to the next. Therefore, our table should give a close approximation for the exact times of spring in the years 2009 and 2010. If you check to learn what these times are, you will find that for 2009 the time is 11:43 A.M. (GMT) on March 20, and for 2010 it is 5:31 P.M. (GMT) also on March 20.

Notice that there are discrepancies of 12 and 13 minutes respectively for these years, compared with our calculations. Why is that? Although the arrival of spring (technically known as the vernal equinox) is defined in terms of the earth and the sun, the moon is a factor. Even if you use a very precise number for the tropical year to calculate from year to year, there may still be a discrepancy of several minutes in any year. However, over the course of many years, the arrival of spring is always within a day or so of March 20.

Israel's Calendar Was Tied to the Seasons

This consistency in the time of spring's arrival is important for our purpose, because the calendar that God gave to ancient Israel is tied to the seasons. We know that God gave them this calendar when they were still in Egypt, as we read in Exodus 12:1-3:

> **And the LORD spake unto Moses and Aaron in the land of Egypt, saying, This month *shall be* unto you the beginning of months: it *shall be* the first month of the year to you. Speak ye unto all the congregation of Israel, saying, In the tenth *day* of this month they shall take to them every man a lamb, according to the house of *their***

fathers, a lamb for an house:

Israel's calendar determined when the feast days were to be observed, but the calendar itself was related to the seasons. For example, in Leviticus 23:39 we read:

> Also in the fifteenth day of the seventh month, when ye have gathered in the fruit of the land, ye shall keep a feast unto the LORD seven days: on the first day *shall be* a sabbath, and on the eighth day *shall be* a sabbath.

As we see in this scripture, the feast of the fifteenth day on the seventh month was kept "when ye have gathered in the fruit of the land." This feast occurred at harvest time. The harvest, of course is determined by the seasons; therefore, we know that their calendar was tied to the seasons.

In Leviticus 23, we are given all the feast days that God commanded ancient Israel to observe. The first one occurs in the first month and the last one in the seventh month. Interestingly, the time of that final feast is elsewhere called the year's end. Exodus 34:22 states:

> And thou shalt observe the feast of weeks, of the firstfruits of wheat harvest, and the feast of ingathering at the year's end.

Does this mean that ancient Israel's calendar had only seven months? No, it doesn't. In 1 Chronicles 27, we read the names of men who served king David month by month. In 1 Chronicles 27:15, we read:

> The twelfth *captain* for the twelfth month *was* Heldai the Netophathite, of Othniel: and in his course *were* twenty and four thousand.

There are twelve names: one for each month; so we know that Israel's calendar had twelve months. In fact, their calendar was based on the cycles of the moon. Scriptures such as Psalm 81:3 point to a lunar calendar:

> Blow up the trumpet in the new moon, in the time appointed, on our solemn feast day.

Appendix 2 Turning Back Our Calendar to the Crucifixion

When we think a bit about a calendar based on the lunar cycle and the seasons, however, we realize that there is a problem. Let's suppose that ancient Israel's calendar began in the spring. The lunar cycle is about 29 and a half days long (29.53059 days to be precise). This means that 12 complete lunar cycles will fit into one year. From the spring of that first year after they had come out of Egypt until about the same time the following year, there were 12 lunar cycles that required a little over 354 days (12 x 29.53059). After those 12 lunar cycles or 12 months had elapsed, they would have begun their next year's calendar, beginning again at the first month. That second year, however, would begin about 11 days earlier than the first year did (365 - 354). If this situation continued from year to year, after a few years the first month - that is, the new year - would begin in the middle of winter. If that happened, the calendar would no longer correspond to the four seasons. The solution was to insert an additional month from time to time, to keep the calendar on track so that the first month always began around spring.

We know that ancient Israel's first month began in the spring from such scriptures as we find in the Book of Joshua. In Joshua 4:19, we read:

> **And the people came up out of Jordan on the tenth *day* of the first month, and encamped in Gilgal, in the east border of Jericho.**

Now read Joshua 3:15:

> **And as they that bare the ark were come unto Jordan, and the feet of the priests that bare the ark were dipped in the brim of the water, (for Jordan overfloweth all his banks all the time of harvest,)**

After spending 40 years wandering through the desert, the children of Israel crossed the Jordan to enter the promised land. The harvest cited in the preceding verse is the first harvest, occurring around spring. It was on the tenth day of the first month, in the spring, that Israel crossed the Jordan.

We know that the first month - known as Nisan or Abib - in Israel's calendar began in the spring. We also know that each month began with the new moon. Ancient Israel's calendar, also known as the Hebrew calendar, is still in use by Jews today (mainly for religious purposes); and it is still based on the lunar cycle.

How can we possibly use all of this information to determine the date of an event that occurred many centuries ago? First, let's look into the way our calendar works. Then we can think about what is meant by saying that we want to learn the date of an ancient event.

About Our Calendar

The calendar we use today is the Gregorian calendar, named after Pope Gregory XIII. It was a reform of the Julian calendar and it dates to February 24, 1582. Even though it was developed over 400 years ago, it wasn't universally adopted at that time. In the British Empire, which included the American colonies before 1776, it wasn't until 1752 that the change was made. It happened on Wednesday, September 2, 1752, which was followed by Thursday, September 14, 1752. In Russia, the new calendar wasn't adopted until 1918.

It is clear that no one who was alive in Biblical times was using our modern calendar. Therefore, when we say that we are trying to establish an exact date for a Biblical event, we mean the following: we want to assign a date to the event such that we would arrive at the present date if the rules of the modern calendar were followed from that past date onwards.

A calendar rule that we need to know is the rule concerning leap years. The Julian calendar inserted a leap year every fourth year. The Gregorian calendar modified that rule as follows:

Every year that is exactly divisible by four is a leap year, except for years that are exactly divisible by 100 (centurial years).

If a centurial year is exactly divisible by 400, then it is still a leap year.

If you are interested in reading more about calendars and have a computer with Internet access, you can consult Wikipedia, the Internet encyclopedia, which supplied this information about the Julian and Gregorian calendars.

The 19 Year Cycle

As man carefully studied the moon over many years, he eventually made a very interesting discovery. We have seen that there are usually twelve new moons in a year, with several days left over. In other words, the duration of the lunar cycle does not fit nicely into a year in an exact number; but over the course of 19 years it fits very well. In fact, man discovered that every 19 years there are 235 lunations - a lunation being the period of time from one new moon to the next. The 19 year cycle is the key that allows us to turn back the calendar.

Our goal is to determine the exact date of the Lord's death, which occurred in 33 A.D. on one of the feast days (Passover) in the Hebrew calendar - a calendar based on the lunar cycle. To do this, we need to find a recent year such that the time interval between that year and 33 A.D. is an integral multiple of 19 year cycles and for which we know the date of the new moon near the first day of spring. Then we can determine the date of the new moon near the first day of spring in 33 A.D. and so count to the date of Passover in that year.

The year 2009 meets our requirement. It's 1,976 years after the year 33 A.D., and that interval is an integral multiple of 19 year cycles (104 x 19 = 1,976). Also, we have accurate information about the dates and times for the new moons in 2009. We know that the first day of spring in 2009 was on March 20. We don't have to know the exact time spring arrived - we only need to know the date. In 2009, the first new moon to occur after spring occurred on March 26. This is the new moon that corresponds to the new moon that began the month of Nisan in 33 A.D.

Does this mean that Nisan 1 in 33 A.D. occurred on March 26? It's not that simple, although March 26 is a good estimate of the date. In order to find the exact date, however, we have more work ahead of us. First, let's take a closer look at the 19 year cycle.

We have learned that there are 235 lunations every 19 years, and that the lunar cycle averages 29.53059 days. This means the following:

235 lunations x 29.53059 days/lunation = 6939.68865 days.

The last of the 235 lunations ends exactly 6939.6885 days after the exact time of the new moon that begins a 19 year cycle. Next, let's see how this number compares with the exact number of days in 19 years as determined by the earth's rotation about the sun. Previously, we used 365 days, 5 hours, 48 minutes, and 45 seconds as the duration of a tropical year. In order to make our calculations easier, we'll use this number in its decimal form (365.2422 days).

19 years x 365.2422 days/year = 6939.60161 days.

Comparing the number of days in 19 years with the number of days required for 235 lunations, we find that there is a difference of 0.08704 days. This is really a very small difference (it's only a little over two hours) considering that we're dealing with a span of 19 years. Nevertheless, over the course of many centuries, we must correct for this difference if we want to find an exact date.

Let's consider the appearance of the new moon that begins the first lunation as an event in time for which we wish to know the anniversary after 19 years. The end of the 235th lunation goes past this anniversary by a little more than two hours. Therefore, if we were using the moon as our clock, we would have to be counting lunations and observing the moon toward the end of its 235th lunation. When the moon was about two hours away from completing lunation number 235, that would be the anniversary of the first new moon that began the 19 year cycle.

Next, let's see what happens 1,976 years after the appearance of the new moon in which we are interested. There are

04 of the 19 year cycles in 1,976 years. Therefore:

104 cycles x 0.08704 days/cycle = 9.05216 days

From this calculation, we see that after 1,976 years the anniversary of the new moon that began the day known as Nisan 1 in 33 A.D. occurred 9.05216 days before the end of the final lunation in the spring of 2009. The number of lunations after 1,976 years is:

104 cycles x 235 lunations/cycle = 24,440 lunations

The moment that ended the final lunation (number 24,440) coincided with the moment that began the new moon in the spring of 2009 (lunation number 24,441, counting from Nisan 1 in 33 A.D.). However, that moment occurred 9.05216 days (plus an exact number of 19 year cycles) after the new moon that began Nisan 1 in 33 A.D. Counting back nine days or so from March 26, 2009 (the date of the first new moon after spring) brings us to March 17. Could it be that Nisan 1 in 33 A.D. occurred on March 17? March 17 is certainly a very good estimate, but we still have to do more calculating before we may know what the date is.

Counting the Leap Years

We have seen that our calendar doesn't correspond exactly with the earth's rotation around the sun. If you want to celebrate an anniversary, undoubtedly you will look at the calendar and celebrate on the same day every year. By doing that, you will be very close to the true anniversary because our calendar works very well. For our purpose, however, we need to make further calculations.

According to our calendar, in a normal year (as opposed to a leap year) there are 365 days as follows:

January	31
February	28
March	31
April	30
May	31
June	30
July	31
August	31
September	30
October	31
November	30
December	31
Total	365

We know that in our calendar, each month consistently has the same number of days, except for February which has 29 days in leap years. When we count a calendar year from a particular day we end up on the same date in the next year. In doing so, we go through 365 days. According to the sun, however, a year takes 365.2422 days. The difference of 0.2422 days over the course of 1,976 years is:

$$1,976 \times 0.2422 \text{ days} = 478.5872 \text{ days}$$

Therefore, if we are interested in a particular event that occurred 1,976 years ago (according to the earth's rotation around the sun), that date would have shifted forward by 478.5872 days if we didn't correct our calendar with leap years. To put it another way, the 1,976th anniversary of January 1 in 33 A.D. is not January 1 in 2009 according to an uncorrected calendar - it's 478.5872 days later: more than a full additional calendar year. Our calendar however, *is* corrected by the insertion of leap years. Let's see how the correction worked over the course of 1,976 years.

The first leap year to occur after 33 A.D. would have been

Appendix 2 Turning Back Our Calendar to the Crucifixion

36 A.D. because the number 36 is an integral multiple of four. After that, there would have been a leap year every four years. The leap years that would have followed 33 A.D. until the end of the first century are as follows:

36	40	44	48	52	56
60	64	68	72	76	80
84	88	92	96		

We don't count the year 100 A.D. as a leap year according to the Gregorian rules. Therefore, there would have been 16 leap years in the first hundred years after 33 A.D.

In the second century, the leap years would have been as follows:

104	108	112	116	120	124
128	132	136	140	144	148
152	156	160	164	168	172
176	180	184	188	192	196

The year 200 A.D. isn't counted as a leap year; so we count 24 leap years for the second 100 years. The third century would also have had 24 leap years; the next hundred years would have one more because the year 400 A.D. would be counted as a leap year. The years 800, 1200, 1600, and 2000 also count as leap years. The centuries in which they appeared therefore have 25 leap years; the other centuries have only 24. The total number of leap years from 33 A.D. to 2009 is summarized as follows:

NUMBER OF LEAP YEARS FROM 33 A.D. TO 2009

33 A.D. to 400 A.D.	89
401 A.D. to 800 A.D.	97
801 A.D. to 1200 A.D.	97
1201 A.D. to 1600 A.D.	97
1601 A.D. to 2000 A.D.	97
2001 A.D. to 2009 A.D.	2
Total	479 leap years

The Doomsday Code

For each leap year, our modern calendar corrects itself by one day. Therefore, we have a forward shift of 478.5872 days when we correct for the tropical year and a backward shift of 479 days due to the leap years. The leap days have actually over-corrected by 0.4128 days (479 - 478.5872 = 0.4128). This means that the anniversary of an event that occurred on a certain day in the spring of 33 A.D. would occur 0.4128 days earlier in 2009. Converting this portion of a day to hours and minutes, we get:

0.4128 days x 24 hours/day = 9.9072 hours
= 9 + 0.9072 hours x 60 minutes/hour = 9 hours, 54 minutes

We are almost ready to use this result in our final calculations, but for now we'll put it aside.

Working Back From the Spring 2009 New Moon

We have seen that the first new moon to occur after the beginning of spring in 2009 occurred on March 26. Now we need to know the exact time of this new moon. The U.S. Naval Observatory (USNO) website has a program that allows us to do this. We are interested in Jerusalem time for this new moon. According to the USNO website, and using Longitude E 35.2 degrees and Latitude N 31.8 degrees as the coordinates for Jerusalem, we find that the new moon occurred on Thursday, March 26 in 2009 at 18:07 (Universal Time + 2 hours). Using the 12 hour convention for stating time, this becomes 6:07 P.M. Jerusalem time.

Earlier, we saw that the time of the March 26, 2009 new moon corresponds to a moment that occurred 9.05216 days earlier than this date in 33 A.D., being the moment that began Nisan 1 in the year the Lord was crucified. Converting the decimal form of 9.05216 days gives us 9 days, 1 hour, 15 minutes. Counting back by this amount of time from the moment of the spring 2009 new moon, we arrive at March 17 at 4:52 P.M. - which is the date and time that the new moon would have occurred in 33 A.D. if our calendar had been perfectly synchronized with the tropical year.

We learned, however, that this is not the case: there was a discrepancy of 9 hours, 54 minutes.

How do we use this 9 hours, 54 minutes to correct our calculation? We saw that the anniversary of an event that occurred in the spring of 33 A.D. would occur 9 hours and 54 minutes earlier than its date in the spring of 2009; or, to put it another way, if we're celebrating the anniversary at a certain time in 2009, then we know that the event occurred 9 hours and 54 minutes later than this in antiquity. Adding the 9 hours and 54 minutes to the time we arrived at earlier (March 17 at 4:52 P.M.), we finally arrive at March 18, 2:46 A.M. as the time of the spring new moon in 33 A.D.

When Was Nisan 1 in 33 A.D.?

Next, we need to realize that Nisan 1 would not begin until *after* the new moon. We know that the Jews began a new day at sundown, at the end of the previous day. One indication of this is found in Leviticus 23:32:

> **It *shall be* unto you a sabbath of rest, and ye shall afflict your souls: in the ninth *day* of the month at even, from even unto even, shall ye celebrate your sabbath.**

This verse, dealing with the Day of Atonement, directed the children of Israel to keep the feast "from even unto even." Another verse that points to sundown as the end of the day is Mark 1:32:

> **And at even, when the sun did set, they brought unto him all that were diseased, and them that were possessed with devils.**

This verse is telling us that people brought to the Lord Jesus those who needed to be healed of various conditions. They came in the evening because - as we learn from the preceding verses in Mark - that day had been a sabbath and they had waited until then for it to end. Although we are accustomed to our days beginning and ending at midnight, it should not be difficult to think in terms of marking the day's end at sundown.

We learned that in 33 A.D., March 18, 2:46 A.M. was the time of spring's new moon. The new month in the Hebrew calendar began immediately after the appearance of the new moon, so it began on March 18 at sunset. On that date, sunset would have been shortly before 6 o'clock in the evening. That's when the month of Nisan or Abib began in 33 A.D.

The Date of the Crucifixion

Counting 13 days from Nisan 1 / March 18 (evening to evening), we arrive at sunset on March 31 as the beginning of Nisan 14. This was the evening the Lord Jesus was arrested by the religious authorities, as we read in John 18:3,

> **Judas then, having received a band *of men* and officers from the chief priests and Pharisees, cometh thither with lanterns and torches and weapons.**

and John 18:12-13:

> **Then the band and the captain and officers of the Jews took Jesus, and bound him, And led him away to Annas first; for he was father in law to Caiaphas, which was the high priest that same year.**

At midnight, our calendar advanced from March 31 to the next day; but according to the Hebrew calendar, it was still Nisan 14. The Lord's ordeal continued all night, through the morning and into the afternoon. That was the day the Lord was crucified: April 1 in 33 A.D., being Nisan 14 (the Passover).

It Was A Friday

There is something else about that day that we can verify. The Bible tells us that the Lord was crucified on a Friday. It doesn't use those words, but based on what the Bible *does* tell us,

Appendix 2 Turning Back Our Calendar to the Crucifixion

we know it was a Friday. For example, the Gospel according to Mark tells us, in Mark 15:42-43:

And now when the even was come, because it was the preparation, that is, the day before the sabbath, Joseph of Arimathaea, an honourable counsellor, which also waited for the kingdom of God, came, and went in boldly unto Pilate, and craved the body of Jesus.

The sabbath at that time was, of course, the seventh day of the week - our Saturday; so the day before it was a Friday.

We can determine which day of the week was April 1 in 33 A.D. by turning back our calendar. We know that March 26, 2009 was a Thursday. Let's see how the day of the week changes for that date as we go back one year at a time.

DAY OF THE WEEK FOR MARCH 26

2008	Wednesday
2007	Monday
2006	Sunday
2005	Saturday
2004	Friday
2003	Wednesday
2002	Tuesday
2001	Monday
2000	Sunday
1999	Friday

Notice how the day changes each year as we go back in time. If we move back one year from a year that isn't a leap year to the previous year, there is a reversal by one day - that is, the date falls on a day that is one day earlier in the week; but if we move back one year from a leap year, there is a reversal of two days.

Earlier, we counted the number of leap years between 33

A.D. and 2009, and we determined that there were 479 of them. We can calculate the total number of days by which the day of a date moves backwards by adding the number of years in the time interval (one day for each year) between the early year and the later year, to the number of leap years (an extra day of reversal for each leap year) between them:

$$1,976 + 479 = 2,455$$

Therefore, there is a shift of 2,455 days on March 26 from that Thursday in 2009 to the corresponding day in 33 A.D. (whatever day it turns out to be). Notice that for each seven days, we have gone through a full week and we end up on a Thursday again. In 2,455 years, there are 350 of these seven-day units:

$$350 \times 7 = 2,450$$

The difference between 2,455 and 2,450 is five. Therefore, the net shift backwards is five days; so we now know that March 26 of 33 A.D. fell on a Saturday. Counting up from Saturday, March 26 in 33 A.D. to April 1 in that year, we find that April 1 was indeed a Friday.

It's important to realize that this *really* was a Friday; that is, it was the sixth day of the week. The date and year number we determined for the Crucifixion depend on our calendar. If the Gregorian calendar worked differently - and it's possible to construct different calendars that work equally well - we could have different year numbers, date numbers and a different name for each month. The days of the week, however, are the same now as they were at Creation. God completed the Creation in one week as we learn in the Book of Genesis. The weekly cycle has been preserved ever since.

This should make us realize that now, whenever a Friday arrives, if we were from that day to go back in time one week at a time, we would eventually arrive at the very day when the Lord died on the cross and completed His work of showing us the penalty for sin; and if we continued going back in time one week at a time, we would eventually arrive at the very day when the Lord created man in His own image.

Similarly, whenever we come to a Saturday, we should remember that it was on that day that the Lord's body rested in the tomb; and it was on that day, being the seventh day, that the Lord rested from all His work of creating during Creation week at the beginning of time.

The Creation began with the day we know as Sunday - the first day of the week. It was on the first day of time that God began His work of Creation, as we read in Genesis 1:3-5:

> **And God said, Let there be light: and there was light. And God saw the light, that** *it was* **good: and God divided the light from the darkness. And God called the light Day, and the darkness he called Night. And the evening and the morning were the first day.**

Just as God created light on the first day of the first week, it was also on the first day of the week that the Lord Jesus rose from the grave. He once again walked among men as the light of the world; and so it is on Sunday that God's elect are to put aside other concerns in order to bring light to the world. Bringing light to the world includes the task of bringing the Gospel message to the world. In our day, that message includes the warning that May 21, 2011 is the last day that God will extend His mercy to save any of us.

Scripture Index

Book

Chapter: Verse   *Page(s)*

Genesis

Chapter: Verse	Page(s)
1:3-5,	339
1:26,	82
1:27,	6, 141
1:27-28,	100
1:30,	100
1:31,	1
2:1-2,	103
2:2,	314
2:15-17,	4
3:6,	76
3:15,	180
3:17,	56
3:23,	57
3:17-19,	4
4:1-2,	181
4:8,	181
4:16,	181
4:25-26,	181
5:2-5,	141
5:3,	314
5:6,	314
5:9,	314
5:11,	315
5:12,	315
5:14,	315
5:15,	316
5:17,	316
5:18,	316

Book

Chapter: Verse   *Page(s)*

Genesis (Continued)

Chapter: Verse	Page(s)
5:20,	316
5:21,	317
5:23,	317
5:24,	182
5:25,	317
5:27,	97, 317
5:28,	318
6:3,	183
6:5,	182
6:5-6,	84
6:8,	182
6:12,	283
6:15,	138, 182
7:1-4,	249
7:4,	184, 261, 288, 319
7:6,	318
7:7-10,	100
7:10-11,	263
7:11,	131
7:11-12,	146, 320
7:16,	250, 321
7:21-22,	250
8:3-4,	132, 259
8:14,	136
8:22,	103, 276
9:29,	96
10:25,	147
11:1,	120
11:4,	120

Book		Book	
Chapter: Verse	*Page(s)*	*Chapter: Verse*	*Page(s)*
Genesis *(Continued)*		**Exodus** *(Continued)*	
11:7-9,	120	12:1-3,	325
11:12,	314	12:3,	101
11:16-17,	142	12:6-7,	102
12:11-13,	26	12:12-13,	102
15:5,	185	12:15-16,	61, 222
17:16-17,	185	12:29-32,	191
19:24,	298	12:35-36,	191
21:5,	186	12:37,	192
22:7-8,	203	12:40,	192
24:59-60,	186	12:40-41,	142
25:29-34,	294	13:15,	290
26:4,	186	13:21-22,	193, 225
35:9-11,	187	14:5-7,	192
35:12,	187	14:9-10,	192
38:7,	290	14:19-20,	226
41:32,	264	14:21-22,	193
45:6,	240	14:27-28,	193
46:2-4,	187	16:29,	64
46:5-7,	188	20:2-17,	44
46:26,	188	20:13,	98
47:11-12,	188	20:18-20,	86
		21:16,	98
Exodus		30:8,	52
		31:1-6,	125
1:13-14,	189	31:13,	52
2:1-2,	189	34:22,	326
3:10,	190	40:34,	125
7:10-12,	127		
8:19,	261	**Leviticus**	
11:1,	191		
12:1-2,	132	13:46,	xxvi

Scripture Index

Book	
Chapter: Verse	Page(s)

Leviticus *(Continued)*

3:46,	xxvi
6:2,	219
7:11,	xxv
9:18,	46
13:10-11,	61
13:15,	224
13:15-16,	62
13:24,	62
13:27,	63
13:32,	335
13:34,	63
13:35-36,	63
13:37-38,	63
13:39,	228, 326
13:42-43,	225
15:3-4,	64
15:6,	65
15:8-9,	65
15:10,	66, 220
15:20-21,	64
15:39-42,	66, 220

Numbers

9:21-23,	225
14:28-30,	281

Deuteronomy

3:28,	195
5:4-5,	46, 73

Book	
Chapter: Verse	Page(s)

Deuteronomy *(Continued)*

6:6-7,	94
10:17-18,	89
11:18-20,	20
16:2-4,	60
16:13-15,	227
24:1-2,	60
25:1-3,	67
34:4-5,	194

Joshua

3:15,	327
4:19,	327
5:6,	148
24:28-29,	149

Judges

2:6-7,	195
2:14,	195

Ruth

3:1,	38

1 Samuel

8:4-5,	196
8:7,	72

Book		Book	
Chapter: Verse	*Page(s)*	*Chapter: Verse*	*Page(s)*
2 Samuel		**1 Chronicles**	
None		16:29,	55
		21:9-10,	134
1 Kings		27:15,	326
2:10-12,	196	**2 Chronicles**	
6:1,	149		
8:12-13,	203	7:1-3,	201
8:66,	244	7:10,	245
10:1-2,	197	14:9-12,	118
11:4,	197	20:3-4,	282
11:11-12,	197	20:12,	282
17:20-22,	291	20:17,	282
19:14,	300	20:22-24,	282
22:6,	306	30:9,	303
22:17,	306	33:9,	198
22:19-23,	306	34:19-21,	85
		36:22-23,	156
2 Kings			
		Ezra	
6:26-31,	280		
18:9-10,	153	6:15,	157
22:6,	19	7:6-8,	158
22:8-11,	19	7:10,	158
23:25,	20		
23:33-34,	155	**Nehemiah**	
24:17,	150		
		None	

Scripture Index

Book

Chapter: Verse   *Page(s)*

Esther

3:7,	159
9:1,	159

Job

1:6-7,	126
23:3-4,	91
38:1-4,	91
38:4,	110
38:17,	110

Psalms

8:3-4,	71
11:5-6,	281
11:6,	319
14:2-3,	76
16:10,	xviii
19:1,	284
21:4,	139
22:14-16,	xvii
22:17-18,	294
27:5,	228
58:3-4,	77
81:3,	326
89:22,	267
90:10,	97
100:5,	279
105:39,	226
107:2-3,	137

Book

Chapter: Verse   *Page(s)*

Psalms *(Continued)*

111:10,	85
118:22,	24
119:11,	311
137:1-5,	199
138:2,	55
148:2,	5
149:2-3,	79

Proverbs

1:27-30,	86
14:12,	76
20:12,	27
31:4,	80

Ecclesiastes

8:5,	41, 288
9:5,	91

Song of Solomon

None

Isaiah

4:1,	212
4:5-6,	226
5:20,	290
6:9-12,	229
6:13,	229

Book			Book		
Chapter: Verse		*Page(s)*	*Chapter: Verse*		*Page(s)*
Isaiah *(Continued)*			***Jeremiah*** *(Continued)*		
7:14,		xvi	32:41-42,		136
8:8,		139	36:1-2,		13
9:3,		34	36:21-24,		17
13:9,		281	36:32,		18
13:13,		276			
14:16,		214	***Lamentations***		
28:10,		32			
34:2-3,		276	None		
40:8,		122			
40:22,		xxv	***Ezekiel***		
40:31,		xxiv			
41:8,		104	1:26,		6
44:28,		199	18:32,		83
44:28 to 45:1		157	31:10-11,		139
46:9-11,		121	33:2-7,		305
52:13		xvi	36:27,		9
53:5-7		xvi	44:21,		80
55:11,		22			
60:2,		242	***Daniel***		
			2:21,		82
Jeremiah			5:23,		3
			8:13-14,		235
1:5,		95	8:14,		136
7:33,		277	9:1-2,		170
8:1-2,		277	9:24,		171
11:3,		56	9:25,		171
17:9,		76	12:8,		xx
18:6,		92	12:8-10,		40, 227
25:11-12,		154	12:9,		xxi
32:9,		135			

Scripture Index

Book Chapter: Verse	Page(s)	Book Chapter: Verse	Page(s)
Daniel *(Continued)*		**Nahum**	
12:11,	236	None	
12:12,	170	**Habakkuk**	
Hosea		None	
13:4,	109	**Zephaniah**	
Joel		2:9,	230
None		**Haggai**	
Amos		None	
3:7,	41, 184	**Zechariah**	
5:2-3,	301	13:8,	237
Obadiah		**Malachi**	
None		None	
Jonah		**Matthew**	
3:4,	288	1:22-25,	xvi
Micah		2:1-2,	xv, 161
5:2,	xv	2:13-15,	162
		2:16,	162
		2:19-21,	163
		3:13,	166

Book		Book	
Chapter: Verse	Page(s)	Chapter: Verse	Page(s)

Matthew *(Continued)*   **Matthew** *(Continued)*

Chapter:Verse	Page(s)	Chapter:Verse	Page(s)
3:13-15,	58	23:23,	222
3:16-17,	166	23:27,	210
4:1-2,	137	24:2,	241
4:4,	xxii, 45	24:11-14,	211
4:18-20,	54	24:14,	23
5:5,	180, 295	24:15-16,	212
5:16,	124	24:21,	231
5:17,	51	24:23-24,	129
5:18,	xxii	24:40-42,	216
5:21-22,	47	25:1-2,	134
5:27-29,	47	27:26,	xviii
6:6,	97	27:50-51,	203
6:9,	310	27:50-53,	277
6:12,	98	28:1,	36, 68
7:22-23,	287	28:2-4,	6
8:21-22,	8		
10:17-20,	14	**Mark**	
10:28,	87		
10:41,	289	1:32,	335
12:39-40,	268	8:15,	222
12:41,	299	10:11-12,	93
12:47-50,	104	12:14,	81
13:10-11,	26	12:15-17,	81
13:27-30,	208	12:29-31,	46
16:12,	222	13:27,	234, 266
18:21-22,	134	13:28-29,	234
19:16-17,	15	13:31,	122
20:31-34,	119	13:32,	266
21:19,	234	15:42-43,	337
22:36-37,	73	15:42-45,	167

Scripture Index

Book

Chapter: Verse　　*Page(s)*

Mark *(Continued)*

15:46-47,	106
16:4-7,	168
16:6,	106
16:15,	123

Luke

1:1-4.	16
1:35-36,	164
2:51,	104
3:1-2,	162
3:16,	27
3:23,	163
3:35-36,	314
4:18-19,	66, 221
10:1,	133
11:21-22,	128
11:42,	24
12:40,	286
12:46-48,	286
13:6-9,	233, 256
13:25-28,	299
14:21-24,	54
15:4,	135
15:10,	6
16:18,	60
17:20,	129
18:9-14,	223
18:10,	87
18:11-13,	88

Book

Chapter: Verse　　*Page(s)*

Luke *(Continued)*

18:13,	99
18:14,	88
19:14,	72
19:27,	72
19:41-42,	84
20:9-16,	202
21:36,	41
22:15-19,	59
22:44,	105, 268
23:9,	xvii
23:42-43,	112
23:44-46,	105, 267

John

1:1,	19, 52
1:1-3,	3
1:12-13,	53
1:29,	101, 167, 221
1:32,	166
1:33,	58
2:19,	237
2:19-20,	199
2:22,	201
3:5-8,	114
3:6-8,	8
3:16,	xxi
3:36,	7
4:12,	16
5:17,	107

Book		Book	
Chapter: Verse	*Page(s)*	*Chapter: Verse*	*Page(s)*
John (Continued)		***Acts***	
5:24,	8	1:4,	204
5:24-29,	278	1:5,	169
6:1-3,	30	1:7,	289
6:44,	xxiv, 99	1:8,	204
6:48-51,	223	1:9-11,	168
6:54,	224	2:1-4,	204
6:54-56,	65	2:31,	xviii
6:63,	22	2:37-38,	205
7:2,	244	2:38,	27
7:37,	244	2:41-42,	205
8:5-9,	83	2:47,	205
8:44,	74	4:4,	206
9:4,	280	4:10-12,	202
10:7-9,	18	4:12,	92
10:17-18,	112	6:7,	207
13:27,	127	6:8,	206
14:1-2,	113	7:57-58,	206
14:15,	55	8:1,	206
16:7,	225	8:27-28,	28
16:13,	29	8:31,	28
17:16-17,	52	8:35,	28
18:3,	336	10:34-35,	90
18:12-13,	336	11:9-10,	39
18:39,	167, 221	11:13-18,	39
19:18-20,	134	11:15,	114
19:23-24,	293	15:39,	215
21:11,	259	16:25-26,	269
21:18-19,	291	17:11,	29
		17:24-25,	3

Scripture Index

Book

Chapter: Verse   Page(s)

Acts *(Continued)*

17:26-27,	118
17:28,	3
27:37,	218
27:41,	217
27:44,	218

Romans

1:18-21,	285
2:13-15,	82
3:10,	289
3:10-11,	xxiii, 77
3:23,	23
5:10,	87, 109
6:23,	178
7:2-3,	93
7:14,	29
7:22-23,	9
8:11,	112
8:28-29	174
9:15,	93
10:17,	92

1 Corinthians

1:26-29,	90, 109
2:12-13,	32
6:9-10,	80
10:16,	59
10:21,	59

Book

Chapter: Verse   Page(s)

1 Corinthians *(Continued)*

15:3-6,	xviii
15:51-52,	179
15:51-53,	115
26:29,	90

2 Corinthians

1:4,	217
5:5,	292
5:8,	10, 292
5:17,	114
5:21,	57
11:13-15,	210

Galatians

3:7,	185
3:13-14,	56
5:2-4,	57
5:19-21,	72
5:22-23,	124

Ephesians

1:3-6,	108
1:4-6,	xxiii, 179
2:-1-3,	74
2:4-7,	10
2:8-9,	xxiv
2:10,	123

Book		**Book**	
Chapter: Verse	*Page(s)*	*Chapter: Verse*	*Page(s)*
Ephesians *(Continued)*		***2 Thessalonians*** *(Cont.)*	
3:10-11,	113	2:1-4,	211
5:19-20,	78	2:3,	213, 267
5:22-24,	94	2:10-12,	307
5:28-29,	94	2:11-12,	249
6:1,	95	3:17,	14
6:4,	94		
		1 Timothy	
Philippians			
		2:1-3,	67
2:9-11,	49	3:15,	25
4:8,	96	5:23,	80
Colossians		***2 Timothy***	
1:12-13,	114	3:1-2,	283
1:16,	5	3:1-5,	210
1:21-22,	87	3:16,	xxii, 31
3:5,	48		
3:24,	295	***Titus***	
1 Thessalonians		2:13-14,	178
4:15-17,	179	***Philemon***	
4:16-17,	251		
5:1-4,	288	None	
5:3-4,	11		
		Hebrews	
2 Thessalonians			
		1:3,	3
1:7-9,	xix	4:3,	110
2:1,	216	4:10,	38

Scripture Index

Book

Chapter: Verse   Page(s)

Hebrews *(Continued)*

4:12,	22
7:1-3,	111
7:26-27,	111
10:24-25,	214
10:31,	87
12:2,	293
12:16-17,	295
12:26-27,	271

James

2:2-4,	89
2:8-10,	89
2:10,	23, 50
2:23,	105

1 Peter

2:9-10,	9
4:7,	Cover page
4:17,	209, 248
5:8,	73

2 Peter

2:5,	183
3:8,	262, 319
3:9,	84
3:10,	10, 178
3:11-12,	251
3:13,	108

Book

Chapter: Verse   Page(s)

1 John

2:15-16,	75
3:4,	50
4:19,	90

2 John

1:1,	90

3 John

None

Jude

17-18,	285

Revelation

1:6,	80
2:7,	208
2:14,	208
3:20,	xxi
4:2-3,	6
7:9,	185, 230, 242
7:14,	230, 243
8:1,	242
9:16-18,	298
12:9,	74
13:8,	110
13:18,	213
14:11,	7

Book

<u>Chapter: Verse   Page(s)</u>

Revelation *(Continued)*

16:14-16,   297
16:18,   271
18:1-3,   213
18:4,   304
20:1-2,   128
20:7-8,   128
21:1,   11, 180
21:2,   172
21:4,   11
21:8,   7
22:18-19,   35